A Study of Magmatic Sulfide Ores

By A. F. Rogers

with an introduction by Kerby Jackson

Introduction

It has been one hundred years since Stanford University released it's important publication "A Study of Magmatic Sulfide Ores". First released in 1914 this important volume has now been out of print for over a century and has been unavailable to the mining community since those days, with the exception of expensive original collector's copies and poorly produced digital editions.

It has often been said that "*gold is where you find it*", but even beginning prospectors understand that their chances for finding something of value in the earth or in the streams of the Golden West are dramatically increased by going back to those places where gold and other minerals were once mined by our forerunners. Despite this, much of the contemporary information on local mining history that is currently available is mostly a result of mere local folklore and persistent rumors of major strikes, the details and facts of which, have long been distorted. Long gone are the old timers and with them, the days of first hand knowledge of the mines of the area and how they operated. Also long gone are most of their notes, their assay reports, their mine maps and personal scrapbooks, along with most of the surveys and reports that were performed for them by private and government geologists. Even published books such as this one are often retired to the local landfill or backyard burn pile by the descendents of those old timers and disappear at an alarming rate. Despite the fact that we live in the so-called "Information Age" where information is supposedly only the push of a button on a keyboard away, true insight into mining properties remains illusive and hard to come by, even to those of us who seek out this sort of information as if our lives depend upon it. Without this type of information readily available to the average independent miner, there is little hope that our metal mining industry will ever recover.

This important volume and others like it, are being presented in their entirety again, in the hope that the average prospector will no longer stumble through the overgrown hills and the tailing strewn creeks without being well informed enough to have a chance to succeed at his ventures.

Kerby Jackson
Josephine County, Oregon
November 2014

CONTENTS

A STUDY OF THE MAGMATIC SULFID ORES

INTRODUCTION

DEFINITION OF MAGMATIC ORES

The term "magmatic ore" is generally applied to those phases of igneous rocks in which there has been an unusual accumulation, supposedly during the molten stage, of the accessory ore-minerals. The recognition of this type of ore deposits is due largely to the work of Vogt, who, in a classic series of papers,[1] not only established the existence of magmatic ores, but also arranged them into well defined groups, and gave the characteristics of each group. His classification has been followed generally by the authors of recent textbooks.[2] Nevertheless some confusion is apparent in geological literature as to the meaning of the term "magmatic" as applied to ore deposits. This is due in part to the designation[3] of all deposits of direct or indirect magmatic origin as magmatic deposits. For example, few doubt the magmatic origin of contact deposits and of cassiterite-tourmaline-quartz veins; but these are not magmatic deposits or segregations in the strict sense of the term.

The term "magmatic deposits" should be limited to those segregations of ore-minerals that take place under the influence of, or closely connected with, the molten stage of the parent rock. Ore accumulations accompanied by destructive pneumatolytic action, or those formed by hydrothermal solutions, are not to be classed as magmatic deposits, altho they may be closely related to, and follow, the magmatic period

---

[1] Vogt, J. H. L.—Bildung von Erzlagerstätten durch Differentiationsprocesse in basischen Eruptivmagmata. Zeit. f. prakt. Geol., Jahrgang 1893, 4-11, 125-43, 257-284.

———Beitrage zur genetischen Classification der durch magmatische Differentiation-processe und der durch Pneumatolyse entstanden Erzvorkommen. Zeit. f. prakt. Geol., Jahrg. 1894, 381-399.

———Weitere Untersuchungen über die Ausscheidungen von Titaneisenerzen in basischen Eruptivgesteine. Zeit. f. prakt. Geol., Jahrg. 1900, 233-242, 370-382; Jahrg. 1901, 9-19, 180-186, 289-296, 327-340.

[2] As is noted below, Vogt classifies as magmatic certain pyritic deposits that we believe should not be included in the magmatic group, and also omits one of the most important divisions of this group, viz., the chalcopyrite-bornite type of magmatic ores.

[3] Geijer, Per.—Iron-ore geology in Sweden and America. Econ. Geol., 10, 231 (1915).

of ore concentration. In as much as ore concentrations connected with persilicic ("acid") rocks are of the latter type, the typical magmatic deposits are confined to the subsilicic ("basic") rocks.

THEORIES AS TO THE SEGREGATION OF MAGMATIC ORES

Altho the chief types of magmatic ores are well recognized, there is considerable difference of opinion as to the details of the processes by which the ores are formed, and as to the "order of crystallization" of the ore-minerals with reference to the silicates. These different ideas may be grouped broadly as follows:

1. Magmatic ores have been defined as unusual accumulations of certain of the accessory minerals of the igneous rocks; and as the accessory minerals are generally believed to be the first to crystallize out of the magma, according to the order of crystallization suggested by Rosenbusch, the natural inference is that the ores are the first to form, and that they settle by gravity to the base of the magma. For example, it is generally assumed that the segregation of the large bodies of iron ore in the Bushveldt laccolith in South Africa and in the Duluth laccolith and the sulfid masses in the Sudbury laccolith have been controlled by gravity.[4]

Beyschlag, Krusch, and Vogt[5] state "Bei den meisten Eruptivgesteinen beginnt die Kristallisation mit der Aussonderung der sogenannten Erzmineralien, wie Magnetit oder Titanomagnetit, Eisenglanz, Ilmenit, Zirkon, Apatit, Schwefelkies, mitunter auch Spinel u. s. w. In einer etwas späteren Stufe der Verfestigung beginnen die Eisenmagnesiumsilikate, wie Glimmer, Hornblend, und Pyroxenmineralien und Olivin, zu kristallisieren."

2. Vogt, who has examined the magmatic ores both in the field and with the microscope, favors the hypothesis that prior to the crystallization of the rock-minerals the ores separate as an immiscible sulfid or oxid melt, which continues in the molten state during the consolidation of the silicates, intrudes the latter, and finally crystallizes.

3. Modern investigation of polished surfaces of ores has led to the discovery that the magmatic sulfid ore-minerals have a definite order of crystallization, and that the younger minerals intrude and replace the older. To meet this, Howe has framed a modification of Vogt's hypothesis. He suggests[6] that the different minerals which make up the sulfid

---

[4] Daly, R. A.—Igneous rocks and their origin, 454 (1914).

[5] Die Erzlagerstätten der Nutzbaren Mineralien und Gesteine, 1, 240 (1910).

[6] Howe, E.—Petrographical notes on the Sudbury nickel deposits. Econ. Geol., 9, 522 (1914).

matte are mutually immiscible, and the sulfids last to crystallize intrude the previously solidified sulfids.

4. Our study of the magmatic ores has led us to frame the hypothesis that *the magmatic ores in general have been introduced at a late magmatic stage as a result of mineralizers, and that the ore-minerals replace the silicates. This replacement, however, differs from that caused by destructive pneumatolytic or hydrothermal processes in that quartz and secondary silicates are not formed at the time the ores are deposited.*

None of the magmatic ores are entirely free from alteration, and in some cases they are associated with high-temperature alteration products as well as with those of hydrothermal origin. It became necessary for us, therefore, to study the various stages of mineral deposition, especially the migrations and alterations that follow the original deposition of the ore.

Some of the data indicating that the magmatic ores are later than the silicates have influenced a number of geologists to question the propriety of classifying certain deposits as magmatic and perhaps to doubt the importance of the type as a whole. Our work meets the objections urged by opponents of the magmatic theories in that it proves the ores are formed within the magmatic period, as defined by us, altho the ore-minerals are later than the silicates.

CLASSIFICATION OF THE MAGMATIC ORES

Vogt[7] separates the magmatic ores according to mineral composition into three groups, viz.: the oxid ores, the sulfid ores, and those of the native metals. The last group includes nickeliferous iron, platinum, copper, and gold. The magmatic deposits of nickeliferous iron and of platinum are curiosities rather than ore deposits. The existence of magmatic copper and gold has been asserted, but not proved by accurate microscopic work. Therefore they are not considered further by us. The oxid group includes deposits of chromite, corundum, and, most important, of magnetite and ilmenite. Altho we touch upon these briefly, the study presented here deals chiefly with the magmatic sulfid deposits. These sulfid ores have been classified as magmatic because of their exclusive occurrence in igneous rocks; because the ore formation is not accompanied by pneumatolytic, contact, or hydrothermal silicates, so markedly developed in connection with sulfid deposits of the non-magmatic types in igneous rocks; and because the

---

[7] Zeit. f. prakt. Geol., Jahrgang 1894, p. 382.

ore-minerals have been considered by some to be contemporaneous with, or even earlier than, the silicate minerals of the igneous rock.

The magmatic sulfid ores have been divided into three groups: (1) the pyrrhotite-chalcopyrite deposits in norite and gabbro; (2) chalcopyrite-bornite deposits in norite and diorite;[a] and (3) the so-called intrusive pyritic ores. We mention the third group but briefly, and doubt the propriety of classifying these deposits either as "intrusive" or as "magmatic."

DEPOSITS STUDIED

Of the first group, the examples studied in detail by us include the ore deposits at Sudbury, Canada; those of the Alexo mine, Canada; and of minor importance, the deposits at Litchfield, Conn.; in San Diego county, California, and of the Golden Curry mine, Montana. We also summarize the important data available in the literature on the deposits of Norway, Sweden, Saxony and Bohemia, and South Africa. The deposits of the second group studied include those of Ookiep, Namaqualand, South Africa, and of the Engels mine, Plumas county, California.

PLAN OF TREATMENT

In the first portion of this article we discuss briefly the conclusions in regard to magmatic differentiation in general that we believe are justified by recent studies by Bowen and others, and summarize our conclusions regarding the magmatic ores, without referring in detail to the microscopic and field data on which they are founded.

In the second part we present our studies of the various sulfid ores examined, and summarize the critical data found in the literature in regard to each occurrence.

Part three contains a brief summary of our conclusions and a statement of the criteria by which the magmatic ores may be recognized.

---

[a] The second subdivision of the magmatic sulfid ores was suggested by Stutzer [Zeit. f. prakt. Geol., 15, 311 (1907)] and the importance of the group was established by the detailed descriptions of the productive deposits in Little Namaqualand [Rogers, A. W.: The nature of the copper deposits of little Namaqualand. Proceed. Geol. Soc. of South Africa, Jan. 31, 1916], and of the Engels mine, California [Turner and Rogers (A. F.): A geologic and microscopic study of a magmatic copper sulfid deposit in Plumas county, California, and its modification by ascending secondary enrichment. Econ. Geol., 9, 359-391 (1914)].

PART I.

DISCUSSION OF THE BEARING OF MAGMATIC DIFFERENTIATION ON ORE SEGREGATION

REVIEW OF THE CURRENT THEORIES OF MAGMATIC DIFFERENTIATION

Magmatic deposits mark the beginning of the ore-forming processes, and they register the early stages of ore formation, often little modified by the complicated processes that follow. The science of ore deposits is largely theoretical on account of the lack of definite information regarding the relation of the ores to the associated minerals. Our microscopic study, carried on during the past two years, is an attempt to get at the facts for this class of ores—a class generally neglected by American geologists.

The conclusions we have reached in regard to the origin of the ores, and to the order of sequence of the ore-minerals, appear to us to have a broader application than to nature's processes of igneous metallurgy. Altho magmatic ore segregations are merely uncommon rock types,[9] nevertheless their study reveals certain of the magmatic processes more clearly and more in detail than do the simpler non-ore-bearing rocks. The relatively new microscopic study of ores may, perchance, make some contribution to the older science of microscopic petrography.

The present time is especially opportune for the presentation of our results because the recent work of Bowen,[10] along both theoretical and experimental lines, affords a foundation in the definite conclusions reached in regard to the important factors governing rock differentiation, and his work possibly has relegated to the scrap heap of discarded theories a number of the hypotheses as to the processes of differentiation.[11]

In view of the detailed discussion by Bowen it is only necessary to mention the two important groups of hypotheses that he now con-

[9] Crook, T.—The genetic classification of rocks and ore deposits. Mineralog. Mag., 17, 55 (1914).

[10] Bowen, N. L.—The later stages of the evolution of the igneous rocks. Jour. Geol., 23 (supplement), 1-91 (1915).

[11] For an excellent summary of the theories of differentiation see L. V. Pirsson, Bull. U. S. Geol. Surv., no. 237, 183-189 (1906).

siders untenable, in order to clear the discussion of the conclusions that are based on them. The first group includes those hypotheses favored by Vogt, Iddings, and Becker that postulate the segregation of mineral compounds in a molten magma by diffusion in a single liquid phase. The flow or diffusion of the materials first to crystallize may be towards the cooler portion of the magma reservoir (the sides) according to the Soret principle, or the segregation may be assisted by convection currents in the magma (Becker), or may be controlled by gravity ("density stratification"). The second group includes the so-called liquation hypotheses, which postulate the formation of immiscible liquid phases (Rosenbusch, Backström, and others). This group of theories has been favored by geologists because it affords a ready explanation of many of the observed field relations, such as the successive intrusion of "basic" and "acid" dikes, etc. However, no evidence of even minute globules indicating immiscible liquids has been found in lavas or in the quenching experiments[12] carried on in the Geophysical Laboratory at Washington.

Theories Favored by Us

We believe that differentiation is accompanied chiefly by sinking of crystals of an early generation to form "basic" rocks, as emphasized by Bowen, and the resultant segregation of "acid extracts" and gases in the liquid portion.[13]

We emphasize the hypothesis that differentiation involves two distinct and at the same time complementary processes: (1) Initial differentiation takes place at an early stage in the consolidation of the magma, at high temperatures, and under relatively anhydrous conditions, and results in the concentration of certain ferromagnesian minerals by the sinking of the early-formed crystals. (2) As a result of the crystallization and removal of the mafic minerals, there is a concentration in the still fluid magma of the felsic constituents, of gases and of mineralizers, of those elements (chiefly the precious and base metals) the crystallization of which is delayed by mineralizers (chlorin, fluorin, boron, water, hydrogen sulfid, etc.). This process finally develops a series of "acid extracts" which form pegmatite and aplite dikes, contact deposits (by the reaction of these extracts upon

---

[12] Bowen, N. L.—Loc. cit., 9-10.

[13] Smyth, C. H. Jr.—The chemical composition of the alkaline rocks and its significance as to their origin. Am. Jour. Sci., (4), 36, 33-46 (1913).

Lane, A. C.—Wet and dry differentiation of igneous rocks. Tufts College Studies (Scientific series), 3, 39-54 (1910).

calcareous rocks), high-temperature quartz veins of the pneumatolytic and allied types, and the various succeeding families of intermediate-temperature ore deposits.[14]

Owing to the vicissitudes of successive differentiation and intrusion, and possibly of remelting and reintrusion, individual masses of "basic" composition are encountered, such as peridotites, gabbros, diorites, and others of "acid" composition such as quartz monzonites, quartz diorites, granites, etc. We believe that, on cooling, each must have undergone an early ("basic") and a late ("acid") differentiation. In the subsilicic rocks one would expect the products of initial differentiation to be most important, and the final "acid extracts" to be small in amount and feeble in action; while the products of the first crystallization of persilicic magmas should not, in general, be rich in the mafic silicates and oxids, but the final extract should develop abundant pegmatite dikes and quartz veins. The latter, altho often important as ore carriers, are not considered magmatic ores in the strict sense of the term, and for this reason we confine our attention to the sulfid and oxid ores segregated in, and at the margins of, gabbro and norite intrusions.

PROBLEMS TO BE INVESTIGATED

Our task is to determine whether these ores, chiefly magnetite, pyrrhotite, chalcopyrite, and bornite, are the early-formed minerals of the magma, or whether they are formed at a late stage, and are accompanied and segregated by the action of an "acid extract" developed by the crystallization of a dominantly "basic" rock. The question is of importance in the general theory of ore deposits. Is ore formation connected with the gaseous extracts developed during the late stages of the consolidation of igneous rocks? Or are there, on the other hand, two unrelated processes of ore formation, one confined to the early stage of the consolidation of the magma, and only of importance in femic rocks, and the other to a later stage, and developed chiefly in connection with salic rocks? If so, the two types of ore will be concentrated at different places, one at the base of the intrusive magma and the other near its upper and outer margins. Their accumulation will be effected by different factors. The correct answer to the question may throw light on the obscure problems of the crystallization of deep-seated magmas. Do all the minerals of igneous rocks crystallize in a regular order according to the laws

---

[14] Tolman, C. F.—The magmatic origin of ore-forming solutions. Min. and Sci. Press, 104, 401-404 (1912).

of solubility in anhydrous melts, or are the relations more complicated? Do the gaseous extracts react on the earlier minerals to form new minerals; and is there in the later stages actual replacement of one substance by another? Are these stages followed by others that are typically hydrothermal? Can these various stages be recognized and distinguished from each other?

METHODS OF INVESTIGATION

These problems may be attacked by three different methods of research, viz.: (1) experimental investigations; (2) field investigations; (3) microscopic investigations.

Experimental Investigations.—Simple rocks consisting of not more than four components can be made and investigated in the laboratory. As yet these have been studied only under anhydrous conditions and without the addition of volatile mineralizers. The work of the staff of the Geophysical Laboratory at Washington has added to the accuracy of our knowledge of crystallization under these conditions. However, both Day [15] and Bowen [16] recognize the importance of the action of mineralizers, but as yet the difficult problem of studying crystallization under the control of volatile mineralizers has not been undertaken. Bowen [17] states: "It will probably be a long time before important aid in attacking the questions can be expected from the experimental side, on account of the difficulty of treating systems containing volatile compounds." Until these are investigated we have not advanced far in the solution, along experimental lines, of the fundamental problems of rock and ore genesis. Up to date experiment has taught us that the common minerals such as olivine, the pyroxenes, and the lime-soda feldspars can segregate by sinking of these crystals in a melt. Analogy suggests, perhaps, that the accessory minerals, especially magnetite and sulfids, which are commonly thought to be among the earliest of the minerals to crystallize out of a magma, might segregate in a similar manner. However, magnetite, apatite, ilmenite, and the sulfids are carried on by mineralizers into the stages of the formation of pegmatites and quartz veins, and therefore are formed at a late stage of rock consolidation. The problem, then, as to whether the magmatic ore deposits of the iron oxids and the metallic sulfids are consolidated during the initial or late magmatic stages has not yet been attacked by experimental work. The experimental researches of Vogt [18] on molten sulfids, valu-

---

[15] Day, A. L.—Some mineral relations from the laboratory viewpoint. Bull. Geol. Soc. Amer., 21, 141-178 (1910).   [16] Loc. cit.   [17] Loc. cit.
[18] Vogt, J. H. L.—Die Silikatschmelzlösungen, (1903-04).

able as they are, have little or no application to the problem, as he experimented with dry melts, and no attempt was made to experiment with an enclosed system into which mineralizers were introduced.

Field Investigations.—The second line of attack is the study and analysis of field relations. Are the magmatic ores always located at the base of a differentiated igneous rock? If this occurs in some cases, is the location at the base of the intrusive rock due to other processes than the sinking of the first-formed crystals? On the other hand, do the ores show any suggestive relation to the salic differentiate (pegmatite dikes, etc.) of the dominantly femic magma? Are the magmatic ores related to fractures developed after partial consolidation, and may these fractures concentrate and release the "acid extracts," and thus localize the ores? Do the ores migrate into, and replace, the country rock? Are the magmatic ores followed in some cases by ore deposits of hydrothermal origin, into which they may grade?

Apparently field evidence alone is not conclusive, or at least has been variously interpreted, for and against, the magmatic origin of the ores. At Sudbury, the location of the ores at the base of the norite sill has been considered strong evidence of magmatic origin during the early stages of consolidation. On the other hand, Knight [19] has recently presented field evidence which he believes is proof of the hydrothermal origin of the ores. The field evidence regarding the deposits investigated appears to us to be of a corroborative rather than of a conclusive nature.

MICROSCOPIC INVESTIGATION

The determination of the age of the ore-minerals in magmatic deposits is merely one phase of the general problem of the determination of the order of crystallization of minerals in igneous rocks. The problem is complex and difficult as far as the early silicates are concerned, as is witnessed by the differences of opinion expressed in recent contributions,[20] and by the doubt[21] thrown on Rosenbusch's fundamental law of decreasing basicity.

These difficulties disappear, however, in regard to the late magmatic minerals, for by microscopic examination of both polished and

---

[19] Knight, C. W.—Origin of Sudbury nickel-copper deposits. Eng. and Min. Jour., 101, 811-812 (1916).

[20] Bowen, N. L.—The order of crystallization in igneous rocks. Jour. Geol., 20, 455-468 (1912).

Ziegler, V.—The order of crystallization in igneous rocks. Jour. Geol., 21, 181-185 (1913). 　　[21] See Vogt, Die Silikatschmelzlösungen, 1, 160.

thin sections (see page 74 for methods employed) we have found evidence that the ore-minerals surround, cut, and replace the earlier silicates.

The more important microscopic investigations of certain of these types of ores include those of Beck, Vogt, Dickson, Campbell and Knight, Howe, and Berg, to which reference is made in the appropriate places. Many of the conclusions of these investigators are verified by us.

Conclusions from our Microscopic Investigations

Some of the important facts that appear to us to be established by microscopic study for both the sulfid and iron oxid ores of magmatic origin are as follows:

The ore-minerals are the final magmatic product, and are formed later than the magmatic hornblende, which we believe to be produced by magmatic alteration.

The ores replace the silicates and, in general, the later-formed ore-minerals replace the earlier ore-minerals.

There is a regular order of formation of the magmatic minerals, which shows no variation in the deposits studied. For the nickel-copper group of sulfid ores it is as follows: (1) silicates, (2) magnetite and ilmenite, (3) pyrrhotite, (4) pentlandite, and (5) chalcopyrite. For the chalcopyrite-bornite group the order is: (1) silicates, (2) magnetite and ilmenite, (3) hematite, (4) pyrrhotite (when present), (5) chalcopyrite and bornite. All alteration minerals, except hornblende, are later than the above mentioned magmatic ores. In some cases minor amounts of "rearranged ores" have been recognized, but the extent of the rearrangement is surprisingly small.

From the field relations we find that the ores may be followed by pegmatite dikes (often containing ore-minerals) and by later series of hydrothermal ore-bearing veins.

The Evidence on which the above Conclusions are based

We conclude that the ores are later than the silicates, for the reason that all the silicates indiscriminately occur as relicts in a ground mass of ore. The ore-minerals surround the silicates, enter along the contacts between them, cut them, and penetrate easily cleavable minerals such as biotite. In some cases they cut the silicates in well defined veinlets. These relations are explained, in part, by those favoring an early magmatic origin of the ores as follows: The sulfid ores remain in a molten condition during the formation of the primary silicates (we add: during the formation of the late magmatic horn-blende), and then solidify.

The presence of hornblende, however, suggests a moderate temperature (far below the melting point of the sulfids) and the presence of water [22] and other mineralizers. Further, it is certain that the ore-minerals have either replaced or corroded the silicates. This is proved by "intersecting structures" [28] and by the fact that portions of crystals have been removed and their place taken by ore-minerals. Berg,[24] describing the magmatic nickel-bearing sulfid ores, states: "Sie (the nickel-bearing sulfids) umschliessen nicht nur gelegentlich alle anderen Gemengteile, sie resorbieren dieselbe nicht nur zu rundlichen Massen, sondern sie korrodieren sie auch öfters, indem sie ganz nach den Gesetz-en der metasomatischen Verdrängung längs Spaltrissen und mechan-ischen Spalten in diese eindringen."

The preservation of the form of antecedent crystals, especially magnetite and olivine, is accomplished by selective replacement. Graphic texture is preserved by the replacement of feldspar of the quartz-feldspar intergrowth. Often the various stages in the replacement of a mineral, from incipient to complete, may be noted.

REPLACEMENT PHENOMENA IN THE MAGMATIC ORES

The process, however, is not one of corrosion, but of replacement. If the ores were molten, corrosion should produce metallic silicates by reaction. No such metal-bearing slag is found. The phenomena are those of ordinary replacement, and the agency that brought in the sulfids removed the dissolved silicates, all of which indicates active mineral-izers.[25]

The regular order in which the sulfid minerals are deposited one after the other, and the fact that one replaces the other, indicates deposition by mineralizing solutions, and not the intrusion of molten sulfids.

---

[22] Harker, A.—The natural history of igneous rocks, 289 (1909).
    Bowen, N. L.—Loc. cit., 41.

[28] Irving, J. D.—Replacement ore bodies and the criteria for their recognition. Econ. Geol., 6, 647 (1911).

[24] Berg. G.—Mikroskopische Untersuchung der Erzlagerstätten, 107-108 (1915).

[25] We regard sulfur as a mineralizer of importance in the magmatic stages in this type of deposits. Sulfur is not usually considered a mineralizer by petrographers, but it is recognized to be such by de Launay and the French school generally, beginning with de Beaumont. [Bull. Soc. Geol. France, 4, pt. 2, 1268 (1847).] The recognition of sulfur as a mineralizer calls attention to the arbitrary distinction between the terms "mineralizer" and "mineralization."

This last point may be met, in part, by Howe's hypothesis that the molten ores are mixtures of mutually immiscible sulfids, and that those last to crystallize penetrate the earlier sulfids. This suggestion does not overcome the difficulties raised by the replacement of the silicates by ores, and antecedent ore-minerals by later ores, without the formation of reaction rims. The absence of the latter shows that the replaced material is removed by the same vehicle that brought in the ore.

The accumulating evidence of the low temperatures [26] at which the final stages of the consolidation of an intrusive magma take place, discredits the notion that a sulfid melt, similar in character to that with which we are familiar in the reverberatory furnace, can exist after the final consolidation of the silicates of the magma. Its "molten" condition must be due to mineralizers, in such amounts that the characteristics of the mixture are those of a high-temperature solution and not of a melt.[27]

We can find no support whatever for the idea that the sulfids separated as molten mixtures and solidified later.

In our microscopic studies of contact-metamorphic deposits we have found definite evidence that the ore-minerals are later than the high-temperature silicates such as garnet, pyroxene, etc. Similar data lead us to believe the same is true for all the high-temperature deposits.

We have come to the conclusion, therefore, that the formation of sulfids takes place at a late stage in all types of high-temperature deposits, probably not higher than 300°-400° C. In this estimate we differ greatly from Lindgren,[28] especially for the magmatic deposits, who believes that the ores are about contemporaneous with the high-temperature silicates.

---

[26] Harker, A.—Loc. cit., 184-188.

[27] Beck calls attention to the physical improbability of molten sulfids entering into the cooler country rock adjacent to the intrusives, and states that the "corrosion" of the silicates has been caused by water solutions, and believes that in many cases the ores have formed after the regional metamorphism of the gabbros, and are younger than the hornblende and garnet. [Lehre von den Erzlagerstätten, dritte Auflage, erster Band, 72-73 (1909).]

[28] Lindgren (Mineral Deposits, 188) gives the ranges of temperature for the high-temperature deposits as follows:

Magmatic deposits, 700° to 1500° C.

Contact deposits and allied veins; pegmatites, 300°± to 800°±.

Vein and replacement deposits formed at great depths, 300°± to 500°±.

ALTERATION AND LATER REARRANGEMENT IN MAGMATIC ORES

The sharp sulfid veinlets will be recognized by all as later than the silicates they cut. They furnish no stronger proof, however, of the late origin of the ores than the larger scale relations, such as the surrounding and penetration of the silicates by the ore-minerals. The well-defined veinlets in the magmatic ores have been explained as "later rearrangements." Lindgren[29] states, in discussing Uglow's description of the replacement and vein phenomena shown in the nickel-bearing sulfids at the Alexo mine, Ontario: "Here, as in so many other cases, secondary changes appear to have been confused with primary deposition." Coleman, writing of the Sudbury ores, states:[30] "That there has been a certain amount of solution and redeposition in many of the ore deposits is admitted by all, but this was of the nature of a rearrangement of the minerals of the rock."

We devoted considerable time to the investigation of the veinlets, vein-like replacements, and alteration products accompanying the ores, in order to determine the extent of these later rearrangements. Numerous sulfid veinlets occur in certain of the Sudbury ores, but the examination of these shows no indication of more than one generation of ore-minerals. These veinlets lead out from large sulfid masses and show no rearrangement, nor are they accompanied by contemporaneous alteration products, as is the insignificant second generation of ore-minerals occasionally met with. Where the sulfid veinlets cut a zone of reticulated fractures filled with alteration minerals, the sulfids do not penetrate these fractures, and therefore they were deposited before the alteration. No greater amount of alteration is found in the ores with, or in the neighborhood of, the veinlets than elsewhere. The sulfid veinlets are found in the normal norite, in the "acid material," and in the "basic" segregations, so that the ore is younger than all these three types of rocks.

None of the magmatic ores are entirely free from the products of secondary alteration, but all the secondary minerals except the late magmatic hornblende are definitely later than the ore-minerals. In a single slide a portion of the ore may be unaccompanied by alteration products, and another portion may be surrounded and cut by later secondary silicates; but the relations of ore to gangue are the same in the altered portions as in the fresh, and there is often no indication

---

[29] Lindgren, W.—Mineral Deposits, 765.

[30] Coleman, A. P.—The nickel industry, Canada Dept. Mines, Report 170, p. 31 (1913).

of secondary migration of ore-minerals, even where they are cut by veinlets of chlorite. The striking thing about the magmatic ores in general is the slight amount of rearrangement they have undergone.

The results of our study of the ores of the Alexo mine, Canada, are of interest in showing that, altho the silicates are intensely altered, and completely serpentinized, nevertheless the ore is only slightly affected. Here the serpentinization is subsequent to ore formation, as shown by veinlets of serpentine cutting the main ore masses. Two generations of serpentinization are recognized; the first accomplished the segregation of insignificant veinlets and specks of chalcopyrite within the area serpentinized, but apparently did not modify even the outline of main masses of ore; the second generation of serpentine is accompanied by still more minute scattered microscopic specks of ore. Only a small portion of the total ore suffered rearrangement during the process of· serpentinization.

ABSENCE OF PNEUMATOLYTIC AND HYDROTHERMAL ALTERATION DURING ORE DEPOSITION

One of the characteristics of magmatic ores that has long been emphasized in the literature is the lack of secondary silicates produced by pneumatolytic and hydrothermal processes. In a few cases, however, the ores are accompanied by high-temperature minerals, such as garnet and tourmaline, and show gradations, therefore, towards the groups of high-temperature epigenetic ore deposits. Our work shows that hydrothermal alteration is invariably later than the period of magmatic ore formation, and therefore emphasizes the lack of alteration during the formation of the ores. It has been argued that if mineralizers or solutions are involved in the formation of the magmatic ores, they must of necessity not only dissolve the rock and deposit the ore, but must also react with the rock-forming minerals to produce secondary silicates. For example, Vogt [81] was at first inclined, from field and microscopic data, to attribute the origin of the nickeliferous sulfids of Norway to "pneumatolytic" action; but the lack of the products of destructive pneumatolysis evidently caused him to abandon the idea that mineralizers are the agents of ore concentration, and he adopted the concept of an intrusive sulfid magma.

The idea that alteration invariably accompanies ore deposition is probably due to a considerable extent to the emphasis given rock alter-

---

[81] [Geol. Fören. Förh, 1883] cited by Beyschlag, Krusch, Vogt, Erzlagerstätten, p. 285.

ation in the classic paper of Lindgren.[82]   As a matter of fact, a careful microscopic examination by us of many types of non-magmatic ores has discovered veinlets or portions of veinlets along which the rock is little altered; and where the rock is affected, this alteration may be earlier or later than the deposition of the sulfids, and not connected directly with it.   In replacement deposits in lime-stone, the most easily altered of all rocks, the ore often lies in contact with the limestone without intervening secondary minerals. *A priori,* no reason can be found why the composition and temperature of the ore-forming solutions in general should not be such that solution of the rock-minerals and deposition of the ore-minerals may take place without the formation of secondary silicates.   *A fortiori,* this would be expected in the case of the ores under discussion.   In the late stages of the consolidation of igneous rocks, constructive action of mineral-izers aids and controls the formation of minerals without the develop-ment of secondary silicates.   We have much to learn from the French scientists of the importance of mineralizers[83] in the crystallization of igneous rocks.

Extensive destructive pneumatolysis often occurs in connection with certain stages in the formation of pegmatite dikes and high-tem-perature veins, and results in the development of such minerals as quartz, muscovite, tourmaline, topaz, scapolite, etc.   The temperature and character of the solutions and mineralizers are such that they are not in equilibrium with the minerals attacked, and they contain chem-ically active gases, such as boron, chlorin, fluorin, etc.   They therefore develop in the country rock the complicated set of silicates mentioned.

---

[82] Lindgren, W.—Metasomatic processes in fissure-veins. Trans. Am. Inst. Min. Eng., **30**, 578-692 (1900).

[83] There is a striking difference between the German and American schools of petrography and the views of the French scientists. The former have inter-preted the conditions governing the crystallization of deep-seated magmas in the light of observations on surface lavas and experiments on anhydrous melts. The latter have been impressed with the fumarolic action around volcanic vents, and have therefore concluded that mineralizers are important components in intrusive magmas. [Lacroix, A.—Les mineraux des fumerolles de l'eruption du Vesuve en avril 1906. Bull. Fr. Soc. Min., **30**, 219-266 (1907).] Extreme views in regard to the role of mineralizers have been expressed by de Launay [Traité de Metallogenie. Gites minéraux et metallifères, 1, chap. 1 (1913)] who has developed a theoret-ical concept of the constitution of the earth involving a central metallic core with a superficial slag or silicate crust. He believes that mineralizers react upon the "slag," producing granite and other rocks, and form various types of ore bodies from the metals they have "extracted" from the metallic core.

We conclude that the magmatic ores, in contrast with the pegmatites and "high-temperature veins," occur in rocks which are little affected by the destructive action accompanying the latter.

The lack of alteration during the formation of the magmatic ores in "basic" rocks is probably the result of chemical composition rather than of temperature. The mineralizers of the "basic" rocks containing the magmatic sulfids produce hornblendization and biotitization, while those of "acid" rocks produce marked destructive effects (greisenization, tourmalinization, silicification, contact action, etc.).

The lack of alteration during ore formation, the fact that ore formation is often followed by the intrusion of pegmatitic differentiates, also the fact that the ores are limited to the parent "basic" rock and migrate into the intruded rock only to a very minor degree, are the chief characteristics of this definite and recognizable type of ore which has been designated as magmatic. Therefore we may well retain this term, even tho certain misconceptions have been attached to it, such as consolidation in the early molten stage, injection of molten sulfids, etc.; recognizing that the microscopic examination definitely proves only that the ores are later than the primary silicates and earlier than the hydrothermal silicates, and also that the period of metallization is further removed from the main stages of rock consolidation than is generally believed.[33a]

MAGNETITE-ILMENITE MAGMATIC ORES

We have studied, chiefly, the magmatic sulfid ores, and for these there is no doubt that the ores are later than the silicates. The same is true for the magnetite-ilmenite ores we have examined; and a review of the literature of these deposits containing accurate microscopic descriptions shows this to be true for all the iron ores of this type, altho the bearing of this fact on the theories of magmatic differentiation does not appear to have been duly emphasized or appreciated.

EUHEDRAL MAGNETITE FORMED AT A LATE MAGMATIC STAGE

Berg [34] recognizes in the majority of magmatic iron-oxid deposits two generations of magnetite: euhedral magnetite earlier than the sili-

[33a] At various times during the progress of our studies, we have considered the advisability of discarding the term "magmatic ore" in favor of some such name as epimagmatic. The fact that we have not been able to discover ores that show evidence of having been formed during the early stages in the consolidation of the magma, that the activity of mineralizers increases as crystallization progresses, and that any division between earlier and later minerals would be arbitrary, has deterred us from suggesting any departure from the present nomenclature, altho such

cates, and larger areas of magnetite later than, and "corroding," the silicates. As magnetite is also a constituent of the magmatic sulfid ores, and as it occurs both as euhedral crystals and anhedral masses replacing the silicates, we were able to investigate the question as to the occurrence of the two generations of magnetite. We find that there are all gradations between small euhedral crystals and the large irregular areas of magnetite. The points of the larger masses may show crystal faces where they penetrate well into the silicate minerals. The magnetite appears to develop irregular forms where it surrounds the silicates, apparently following the lines of least resistance along the mineral boundaries, but develops euhedral forms within the silicates, where no line of weakness disturbs its tendency to crystallize regularly. We believe that the magnetite in the magmatic sulfid ores, and probably in the magmatic ores in general, is all deposited later than the silicates.

OTHER ACCESSORY MINERALS OF PROBABLE LATE MAGMATIC ORIGIN

The accessory minerals of igneous rocks, such as magnetite, ilmenite, apatite, titanite, zircon, etc., are considered by petrographers to be the first to crystallize in igneous rocks, on account of their euhedral form, and their occurrence within the silicates that make up the mass of the rock. As the effect of mineralizers is not generally recognized in developing new minerals in the igneous rocks the possibility has not been considered that these minerals form in the later magmatic stages by replacement.[35] Euhedral crystals of pyrite in igneous and sedimentary rocks and of garnet and magnetite in metamorphic rocks, are found unaccompanied by any adjacent alteration zone. In a similar manner we believe that euhedral magnetite, and probably other accessory minerals, are the product of mineralizers developed during the late magmatic stage.

The high-temperature minerals forming the bulk of the igneous rocks are formed at an early stage. The accessory minerals present in small amounts are not formed until a late stage and then under the influence of mineralizers. The difficulty is thus avoided as to how the magma, at an early stage, could become saturated in relatively insoluble compounds present in small amounts, and how these again could make their appearance in large amounts among the last products of consolidation.

a change might assist in clearing up misconceptions in regard to the magmatic ores and the processes of magmatic differentiation.   [34] Loc. cit., 102, 105.

[35] "If crystals of one primary mineral completely enclose crystals of another, the evidence of their relative age is absolute." (Harker, loc. cit., 179.)

Stages Recognized in the Formation of Magmatic Ores

From the above discussion it is clear that we conceive of the process of formation of plutonic rocks as consisting of stages, and that rock differentiation and ore formation are the results of an orderly series of events.

The process varies in detail with the composition and size of the individual intrusive masses undergoing crystallization. The stages in the norites and gabbros which contain the magmatic sulfid ores are as follows:

(1). The first minerals to form are olivine, the pyroxenes, and the feldspars.

(2). Magmatic alteration of the silicates often takes place prior to the formation of the ore-minerals. The most common change is that of pyroxene to hornblende (not uralite). The not uncommon hornblende gabbro, for example, may be developed by this late magmatic process, for the hornblende has probably been formed at the expense of pyroxene.

(3). Later magmatic products include interstitial pegmatitic material, interstitial quartz, and occasionally tourmaline, garnet, analcite, epidote, and calcite.

(4). The introduction of the ores by mineralizers is later, in general, than the minerals of group (3) and is unaccompanied by any secondary silicates.

(5). The pegmatite dikes, found in the neighborhood of almost all of the magmatic sulfid ores, are often later than the magmatic deposits of the basic rock itself.

(6). Hydrothermal alteration subsequent to magmatic ore deposition includes the development of chlorite, tremolite, anthophyllite, sericite, and serpentine. In general, hydrothermal alteration, altho seldom lacking, is insignificant compared with that developed in connection with deposits of other types in igneous rocks, such as those of Butte, Bingham, etc. It often does not accomplish any rearrangement of the ore, altho, in some cases, insignificant amounts of pentlandite, chalcopyrite, chalcocite and covellite are formed.

(7). At a later stage, downward enrichment and oxidation may take place. Magmatic deposits, in all cases examined, owe their metallic content to the original magmatic minerals and not to later introduced sulfids.

PART II.

DESCRIPTION OF THE VARIOUS DEPOSITS OF THE MAGMATIC ORES

GROUP I. THE NICKEL- AND COPPER-BEARING PYRRHOTITIC DEPOSITS

SUDBURY, CANADA

GEOLOGY

Those who favor the hypothesis that the Sudbury ores separated out of the magma as an early, or the earliest, constituent, and sank and collected by gravitative differentiation,[86] cite field relations as proving their contention. In spite of the voluminous literature in which these relations have been described, there are still many points in regard to which no agreement has been reached. A discussion of the genesis of these deposits is therefore incomplete without a summary of the established field relations and a statement of the problems that are unsolved as yet. Fortunately Coleman [87] has mapped in detail both the upper and lower margins of the laccolithic sheet, and given much information in regard to the structure of the individual deposits. The accompanying map (fig. 1) is drawn from a photographic reduction of his large-scale map.

As is well known, the pyrrhotite-pentlandite-chalcopyrite ore bodies of the Sudbury district occur at the base of a great sheet of "norite-micropegmatite." As shown on the map, the sheet is of unusual regularity for an intrusive mass of its size and character. It is overlain by 9000 feet of Upper Huronian sandstones, shales, conglomerates, and tuffs, which make a gentle syncline concordant in dip with the norite sheet, and underlain by a sedimentary series of great thickness (30,000 feet according to Coleman), predominantly of quartzite with included older basic irruptives. These older strata, "the Sudbury series" of early Huronian or pre-Huronian age, dip regularly to the southeast, averaging

---

[86] Coleman, A. P.—The nickel industry. Can. Dept. of Mines, Mines Branch. Publ. 170 (1913).

Barlow, A. E.—Geol. Surv. Can., 14, pt. H, 1-124 (1904).

[87] Loc. cit.

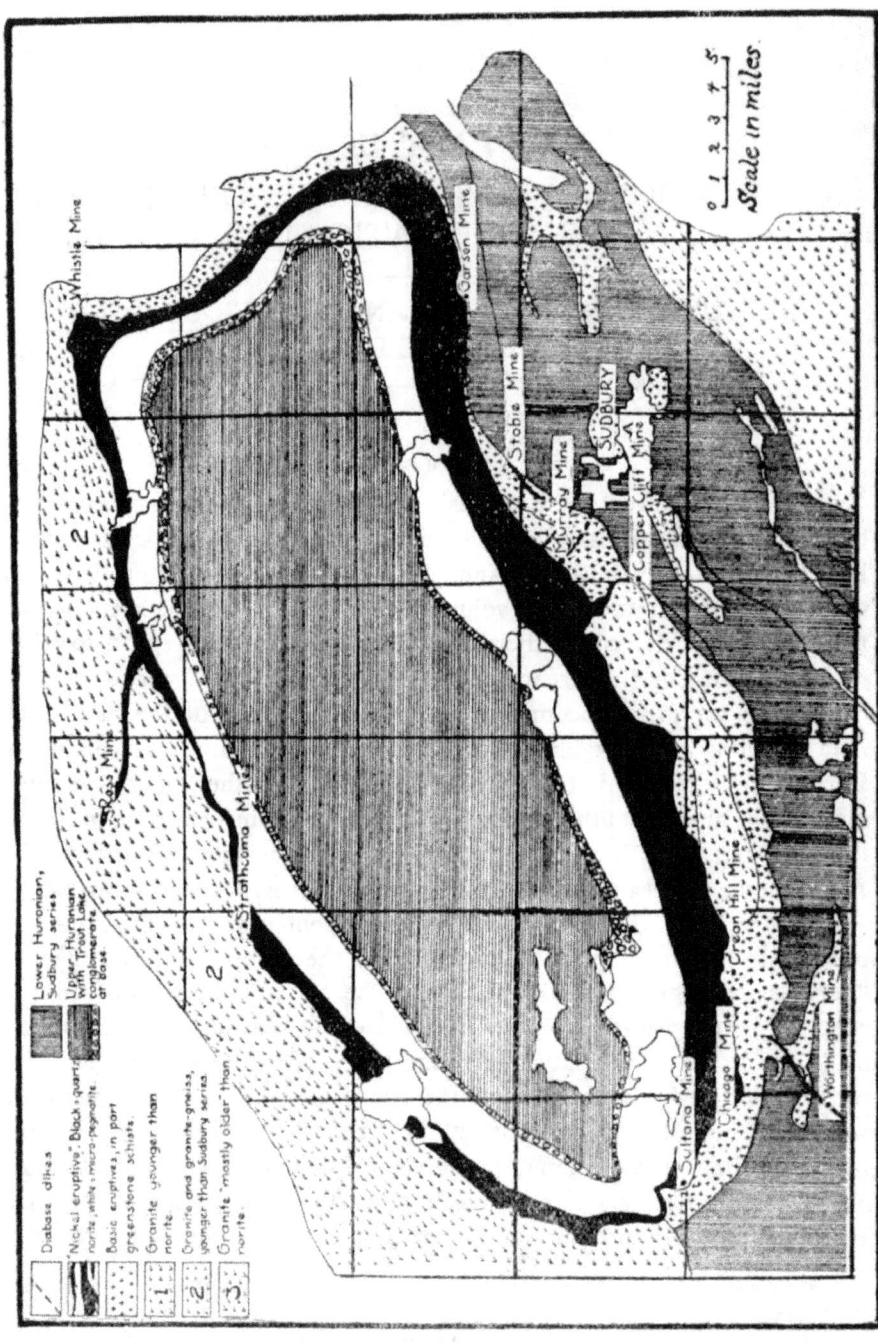

Fig. 1. Geologic Map of the Sudbury District. Drawn from a photographic reproduction of Coleman's latest map accompanying The Nickel Industry, Publ. No. 170, Canada Dept. of Mines.

45°. The "nickel eruptive," as the norite sheet has been named by Coleman,[88] therefore, has taken advantage of the unconformity between the older and younger sedimentary series, and has lifted the latter on its back without disturbing it otherwise than to develop a gentle spoon-shaped syncline.

No accurate sections across the syncline have been made, but judging from the areal map and the descriptions and sketches of the ore bodies located at the contact between the nickel eruptive and the underlying formations, the dip of the eruptive sheet at its basal contact is 40° to 45° at the surface, generally decreasing to 30° to 35° at a depth of about 1000 feet.

Coleman [89] has pictured in the following vivid language his concept of the manner in which the great laccolith forced its way into the sedimentary rocks:

"After the succession of sediments just mentioned had been deposited, the vast mass of molten rock of the nickel eruptive ascended, mostly from beneath an area near the middle of the southern range, as will be shown later. As the molten magma welled up from below, the crystalline rocks forming the roof of the great crucible gradually collapsed as a block twelve to fifteen miles long and several miles broad, giving rise to extensive faulting and fissuring. . . . "

"The eruptive sheet cooled extremely slowly, partly because of the great bulk of molten material and partly because of the thick mantle of sedimentary rocks above; and during the cooling much of the ore sank to the bottom, though its upper part remained mixed with norite, which finally blended into micropegmatite or granite on top, the three materials arranging themselves according to their specific gravities. . . . "

"The norite-micropegmatite sheet is one of the largest laccolithic sheets known, containing not less than 600 cubic miles at the present time, and probably having had a much greater bulk in the beginning."

Subsequent to the intrusion of the nickel eruptive, both granite masses and dikes and basic dikes were intruded along the under contact of the sill, and both "acid" and "basic" dikes cut the ore bodies.

One of the most important problems, as yet unsolved, is the origin of the "acid material," largely quartz and feldspar, frequently developing marked micropegmatitic structure, that occurs enclosed in many of the

---

[88] Coleman, A. P.—The Sudbury laccolithic sheet. Jour. Geol., 15, 752-782 (1907).

[89] Canadian Dept. Mines, Pub. 170, p. 10 (1913).

ore bodies. Equally important is the determination of the relation of the "later granite" to the norite. Coleman has arranged the granitic rocks in three groups (see geological map, page 24). (1) Granite and granite-gneiss younger than (intrusive into) the Sudbury series and older than the norite; (2) granite mostly older than norite which was mapped without discrimination between "the older" and "the younger" groups which are of widely different origin; (3) granite younger than norite, a small area of which is shown near the Murray mine. The lack of detailed field and microscopic study prevents, for the present, the final answer to the questions raised above. Fortunately, however, Knight[40] has been studying these questions in the course of his detailed field work, and we look forward to the publication of his conclusions with great interest.

In regard to the micrographic segregations and inclusions of granitic material in the ore, we believe with Coleman[41] that they are differentiates of the norite.

The micrographic structure is typical of the upper portion of the "nickel intrusive." Its occurrence in the ores as blebs and interstitial material, as well as large masses, and the variations shown in a single thin section, suggest that both the norite and the "acid" material are differentiates of a common magma.

In this regard Howe[42] states:

"The attractive possibility has been considered that the silicious material associated with the sulphides might represent a residual portion of the magma from which the sulphides are supposed to have been derived. . . . The microscope, however, shows over and over again the absolute similarity of the acid inclusions to the foot-wall granite."

However, the foot-wall granite to which he refers may well be the "later granite." Specimens sent us by Dr. Knight from the Creighton mine, the locality which Howe studied, are decidedly micrographic. Until further data are available, the reasonable hypothesis is that the "acid" blebs, segregates and inclusions in the Sudbury ore bodies, as well as the marginal masses of the later granite, are differentiates of the parent norite magma. In many other localities, granitic differentiates are found inclosed in, or at the margins of, the norite intrusives (see the map of the Ringerike district, Norway, page 49). At Sudbury, where all the phenomena have taken place on a large scale,

[40] Loc. cit. This article contains a preliminary statement of certain of the field relations.   [41] Econ. Geol., 10, 391-392 (1915).   [42] Loc. cit., 520-521.

we should expect the "acid" differentiates to appear in large amounts. The fact brought out by Knight that dikes of the later granite cut the norite, and also that the ore rests upon a footwall of younger granite, does not prove that the ores are not magmatic. The norite and granite differentiates may be nearly contemporaneous, and the "late magmatic" ores are, in some localities, earlier than, and in other places later than, the "acid" differentiates. Evidence of extensive differentiation is found generally in rocks associated with magmatic ores.

Knight, [42a] has also emphasized that the Sudbury ores have wandered out into, and replaced, the schists found at the contact with the norite.

We are indebted to Dr. C. W. Knight for specimens of several types of sulfid bearing "greenstone schists." These laminated rocks consist largely of biotite, hornblende, actinolite, chlorite, clinozoisite, quartz, and plagioclase. One of these, a chlorite-actinolite gneiss from the Garson mine, is shown in fig. 10. The ore-minerals, pyrrhotite and chalcopyrite, occur in linear areas parallel to the plane of schistosity. Under the microscope the relations of the ore-minerals to the silcates are practically the same as in the igneous rocks examined. The sulfids surround the silicates, and evidently replace them to some extent, and minute sulfid veinlets definitely cut the silicates. Alteration products are practically absent.

This shows that the magmatic ores migrate to some extent into the country rocks, and apparently show the same replacement phenomena as are exhibited by the ores in the igneous rocks.

STRUCTURE OF THE ORE BODIES

Coleman, in his recent description of the Sudbury region, recognizes the following types of ore bodies:

1. *Marginal Deposits.*—These are sulfid segregations collected at the base of the main norite sheet. They gradually fade out above into barren country rock, and often show brecciation, fissuring, and "later reconcentration of ores", especially along the foot-wall, which is always a pronounced fissure. *Faulted marginal deposits* have suffered brecciation and faulting in their upper portions, and the ore has "wandered into the fissures between the blocks, either at the time as molten sulphides, or later through water transport. As chalcopyrite is everywhere the more transferable of the sulphides, it has entered the fissures more largely than the pyrrhotite. Unusually large amounts of quartz, carbonates, and

---

[42a] Loc. cit.

sulphides of zinc or lead are found in these two mines (Crean Hill and Garson mines) as a result of circulating waters, and these later processes have played a larger part than in most other ore bodies of the region. whether marginal or offset."[43]

2. *Columnar Offset Deposits.*—These remarkable ore deposits, the most notable of which occur in the Copper Cliff mine, are great cylindrical ore shoots.

"In the last report on the nickel region the Copper Cliff deposit was known to go down for 1,000 feet without interruption, as a rude oval pipe with diameters varying from 50 to 200 feet, and a dip of 77½° to the northeast.

"Since that time the two ore bodies of the Victoria mine, though smaller in diameter, have been followed to the depth of 1,400 feet with no indication that they may not continue indefinitely. These two small cylinders of ore, more than 1,400 feet in length and close together, but never meeting, are not at all easy to account for on any other theory than the magmatic one, and this continuance to so great a depth was not anticipated in earlier studies of the region. . . .

"There is usually more evidence of water action than in the marginal mines, and often a certain amount of quartz and of rusty weathering carbonates is mixed with the ore, probably as later effects of magmatic waters." [44]

The following ingenious explanation of these columnar ore shoots as dikes or apophyses running out from the main laccolith is favored by Coleman: ". . . it is possible that the most fluid part of the magma, the pyrrhotite-norite, entering all the fissures produced by the collapse of the underlying rock, rose from beneath under hydraulic pressure and was able, in a sense, to drill holes up through the crushed zones of rock above." [45]

3. *Parallel Offsets.*—These include the great Frood-Stobie ore deposit described by Coleman as follows:

"The Frood-Stobie offset runs nearly parallel to the basic edge, but at a distance of from ¾ of a mile to 1½ miles to the southeast. The ore more nearly resembles that of a marginal deposit than that of the ordinary offsets; and the ore body dips at an angle of 60° toward the basic edge. It is a long irregular sheet enclosing much rock, and its connexion with the edge of the norite is probably at a considerable depth below the surface. The margin of the norite parallel to it shows comparatively little ore, the sulphides belonging to it having been drained

---

[43] Loc. cit., 35.   [44] Coleman.—Loc. cit., 36-37.   [45] Loc. cit., 37.

off through a complex set of fissures to the Frood-Stobie deposit. The ore is known by diamond drilling to extend northwest beneath the country rocks to a depth of more than 1,000 feet, and at the lower points it distinctly flattens toward the basic edge of the norite. No other deposit of this type has so far been discovered; but the Frood-Stobie belt of ore is so important and so very distinct from the other types that it deserves a place by itself."

From the above descriptions it appears that the parallel offsets are mineralized dikes or sills parallel to the main foot-wall contact and dipping toward it.

From the quotations given above, it will be seen that there are evidences of hydrothermal deposition of ore and gangue minerals distinctly later than the main ore mass of the various types cited. Pyrite, marcasite, galena, sphalerite, and molybdenite are found in later veinlets, often cutting the main ore-bodies and accompanied by quartz and calcite.

MICROSCOPIC DESCRIPTIONS

We are greatly indebted to Messrs. F. L. Hess, J. F. Kemp, C. W. Knight, R. D. Longyear, F. H. Mason, M. E. Morgan, H. Ries, and T. L. Walker for specimens of rocks and ores from various mines of the Sudbury district. Our study has been facilitated by the excellent suite of Sudbury rocks obtained from the Royal Ontario Museum of Mineralogy. Our specimens are believed to be typical of the Sudbury ores, as we repeatedly find the structures and relations described by other workers.

There are three fairly distinct rock types directly associated with the ores studied by us: (1) quartz norite almost free from sulfids; (2) pyrrhotite norite with fair amounts of the sulfids and with uralitized pyroxene and some hornblende; and (3) a hornblende-bearing granitic rock with abundant sulfids constituting "the rich ore." These three types represent in a general way the three kinds of material found at the mines: (1) the "lean ore" or the barren norite, (2) the pyrrhotite norite transitional to the ore, and (3) the massive ore.

Lean Ore from the Stobie Mine.—One of the lean ores studied in detail by us is a rather fine-grained quartz norite, consisting of hypersthene, plagioclase, quartz, subordinate biotite and hornblende, magnetite, pyrrhotite, chalcopyrite, and smaller amounts of secondary chlorite and tremolite. The general relations are shown in fig. 11 (plate II).

The magnetite occurs in euhedral, subhedral, and anhedral crystals. Most petrographers would assign the magnetite to the first period of crystallization, but there is clear evidence that the anhedral magnetite was formed later than the silicates. This is shown clearly in fig. 12. The magnetite has completely surrounded one hypersthene crystal and partly surrounded two others. All of the magnetite belongs to one generation, for there is a perfect gradation from the euhedral to anhedral forms. Thus we have evidence that the euhedral magnetite was formed during the late magmatic stage. The sulfids, pyrrhotite and chalcopyrite, which are found in occasional spots, are also later than the silicates, for with the magnetite they form hook-shaped anhedra surrounding the silicates, as shown in fig. 12. The ore-minerals occur between the silicate anhedra, and while they often surround the silicates they rarely cut across an individual crystal. Careful search, however, usually reveals a few occurrences of this sort.

Alteration products occur in the sections figured, and the natural inference would be that the sulfids are connected in some way with the alteration. Alteration of the hypersthene to tremolite and the formation of chlorite have taken place, but these minerals have been formed after the introduction of the sulfids, as figs. 13 and 14 prove. In fig. 13 a chlorite veinlet cuts pyrrhotite and chalcopyrite, and in the polished section of fig. 15 similar relations are shown. A veinlet of sulfids to the left of the chlorite veinlet of fig. 13 is accompanied by tremolite, but, as shown in figs. 27, 28, 29 and 30 (plate VI), the sulfid veinlets have no connection with alteration products. The hypersthene in the lower part of fig. 12 is partially altered, but it shows exactly the same relation to the ore-minerals as does the unaltered hypersthene in the upper part of the figure. Fig. 14 furnishes evidence that the tremolite is later than the ore-minerals. The sharp needle-crystals are not residual, but project out from the altered hypersthene into the chalcopyrite mass.

The quartz norite lean ore from the Stobie mine affords clear evidence that the ore-minerals were formed at the end of the magmatic stage, that the slight alteration of the hypersthene took place after the introduction of the ore-minerals, and that none of the sulfids have undergone rearrangement of any kind.

Pyrrhotite Norite from the Stobie Mine.—We have studied a typical specimen of pyrrhotite norite from the Stobie mine. This rock constitutes a medium grade ore with large crystals of plagioclase, aggregates of tremolite, rims of hornblende, and sulfid masses. The general relations are shown in fig. 16. The light gray areas represented in the photograph are largely aggregates of uralite needles in more or less parallel position. These aggregates are often surrounded by hornblende. The probable explanation is that hypersthene was bordered by rims of hornblende, and at a later stage the hypersthene, but not the hornblende, underwent uralitization. From other data we know that the sulfids were formed after hypersthene and hornblende. A study of polished sections proves that the uralitization (tremolitization) occurred after the introduction of the sulfids, and as evidence we introduce figs. 17 and 18.

Fig. 18 shows pentlandite of the first generation in veinlike areas in the pyrrhotite, and pentlandite of the second generation developing along crystallographic directions of the pyrrhotite. The rearrangement of the ores is a very minor feature in this, as in the other Sudbury specimens.

The Rich Ores.—The massive ores examined by us show residual spots of rather "acid" material, consisting largely of alkali feldspar, quartz, hornblende, and biotite. In fact the gangue of the rich ore is more like the so-called micropegmatite than the quartz norite. Microcline is abundant, and is often intergrown with quartz. In none of the rich ores have we been able to find hypersthene. Its place seems to be taken by hornblende and biotite, which are probably late magmatic alteration products. It is difficult to believe that the "acid" rock is an older foot-wall granite. It is more probably a felsic differentiate of the same magma that furnished the norite.

Rich Ore from the Stobie Mine.—Rich ore from the Stobie mine is represented by fig. 19 (polished section). The chief minerals are plagioclase, quartz, and biotite. Garnet is also present. The ore-minerals, which occur in irregular masses occasionally extending out into veinlets, replace all the silicates including garnet. The replacement along cleavage lines of biotite by the chalcopyrite is beautifully shown in thin sections. This specimen is almost entirely free from alteration products. There is, however, a little chlorite in small lath-shaped sections, and these distinctly cut the ores.

Rich Ore from the Copper Cliff Mine.—On plate I (fig. 7) we show a large polished hand-specimen from this mine. This specimen shows silicates which are believed to be residual. The residual spots contain microcline and quartz (often in graphic intergrowth), plagioclase, biotite, a little hornblende, and long acicular crystals of apatite. As far as can be told from the residual matter which has escaped replacement, the rock is a granite. The ore-minerals, magnetite and pyrrhotite, replace the silicates, especially biotite, as is shown in fig. 20. The biotite gives one the impression of being one of the last formed silicates. There are also very small biotite crystals, which may possibly belong to a second generation. The apatite is also formed at a late stage.

There are a few alteration products present, such as chlorite and sericite, but they have nothing to do with the ores.

Rich Ores from the Creighton Mine.—The Creighton is the largest mine in the Sudbury district, and fortunately our suite of specimens includes a number from that deposit. The residual silicates in the ores are plagioclase, microcline, hornblende, and biotite. Hypersthene is lacking. Quartz is abundant as an interstitial mineral, and often occurs as a quartz-feldspar intergrowth.

On account of the abundance of quartz, one of the rich ores is practically a granite or possibly a grano-diorite. Another specimen very much resembles the quartz norite in structure, but hornblende is present instead of hypersthene. It is probable that this rock was originally a norite, and this indicates the possibility not only of late magmatic minerals but also of late magmatic rocks. The same specimen that furnished the thin section showing the norite texture also shows microcline and a subgraphic intergrowth of quartz and feldspar. This variation in the constituents within a small space is a characteristic of the igneous rocks containing magmatic ores.

A photograph of rich ore from the Creighton mine is shown in fig. 8 (plate I). This specimen furnished us the photomicrographs of figs. 25, 27-30, 32, and 33. A similar specimen furnished photomicrographs 23, 24, 31, and 34.

Polished sections showing the general relation of the sulfids to the silicates in these specimens are shown in figs. 27, 28, 31, and 32. Thin sections, which are necessary for the identification of the silicates, are represented by figs. 23 and 25.

The sulfids are later than the silicates. This is almost certain from the irregular hook-shaped anhedra which surround and occasionally project into the silicates; but if any doubt exists as to the later origin of the sulfids note the veinlets in plates V, VI, and VII. The veinlets replace almost indiscriminately the various silicates (see figs. 24, 28, and 30), but certain sulfid areas show marked selective replacement of the feldspar of the quartz-feldspar intergrowth, as in fig. 31. For an especially good illustration of replacement veinlet see fig. 24, where fresh biotite has been invaded by the sulfid. In fig. 26 a hornblende crystal has been cut squarely in two by a sulfid veinlet. Note the branching veinlets of figs. 28 and 30. The only explanation of these veinlets is that they are later than the silicates. The proponents of the magmatic theory for the Sudbury ores have ascribed these veinlets to rearrangement, and thus they reconcile their theories with microscopic work such as that of Dickson. That the sulfid veinlets are of the same generation as the main sulfid masses and are not due to later rearrangement is shown in fig. 29, where the sulfids extend out into a veinlet without any break in the continuity or other evidence of later origin.

The late origin of the sulfid veinlets can hardly be doubted, but some may be inclined to suggest a hydrothermal origin for them. This is disproved by our findings. For example in the thin sections of the Creighton ore we found tremolite (and possibly talc) pseudomorphous after original hypersthene. The pseudomorphs contain minute magnetite crystals which were formed by alteration. The pseudomorphs are cut by veinlets of chalcopyrite, but the chalcopyrite has not wandered into the cracks of the tremolite. This shows that the hydrothermal alteration of hypersthene to tremolite (or talc) and magnetite occurred after the introduction of the chalcopyrite.

That the magnetite, as well as the sulfids, is formed at a late magmatic stage is indicated by fig. 20. The magnetite contains included ilmenite plates as illustrated by fig. 21. This magnetite-ilmenite intergrowth is a characteristic feature of magmatic ores.

Polished sections of massive, almost solid, ore from the Creighton mine are represented by figs. 22, 35, and 36. Fig. 22 shows an area of silicates which is doubtless a relict of the same granitic material present in the other Creighton samples. Magnetite has been replaced by the sulfids and the pentlandite has replaced pyrrhotite in vein-like masses. Veins of pentlandite are well shown in fig. 35; also minute tufts of brush-like crystals of a pale-yellow mineral, probably pentlandite of a second generation. Fig. 36 is a highly-magnified view of one of these crystals which extends out from a veinlet of chalcopyrite, probably of the second generation. Figs. 35 and 18 show what a minor amount of rearrangement has taken place in the typical magmatic ore of Sudbury.

At some of the Sudbury mines, notably the Worthington, certain sulfids which are not typically magmatic have been found. Among these minerals are pyrite and polydymite, but many others have been reported.[46]

We have examined several Sudbury ores containing pyrite. In fig. 37 is represented a supposed specimen of polydymite from the Vermilion mine. This contains pyrite in the form of veinlets evidently formed at a late stage. The relation of the polydymite to the magmatic sulfids is not entirely certain, but the examination of a very fresh speci-

---

[46] Barlow, A. E.—The nickel and copper deposits of Sudbury, Ontario. Geol. Surv. Can. Ann. Rept. 14, pt. H, 93 et seq.

men suggests that the Sudbury polydymite is a mixture of three minerals: pentlandite, an unknown violet-gray mineral, and the true polydymite. The polydymite and violet-gray mineral are probably due to the breaking down of the pentlandite.

One of the later sulfid ores from the Worthington mine is shown in fig. 38. Pyrite and sphalerite occur in a gangue of calcite. The pyrite has a peculiar reticulate structure. The sphalerite is later than pyrite.

The order of succession of the ore-minerals at Sudbury as determined in polished sections is as follows: (1) magnetite, (2) pyrrhotite, (3) pentlandite, and (4) chalcopyrite. The other sulfids, such as pyrite, sphalerite, etc., are post-magmatic.

ORIGIN OF THE SUDBURY ORES

With the possible exception of the gold-bearing "banket" of the Rand (South Africa), perhaps no other single group of ore deposits has received as much attention from the standpoint of origin as have the Sudbury deposits. Both the igneous and hydrothermal hypotheses of origin have had an almost equal number of advocates. Altho the earliest papers on the Sudbury deposits advocated the aqueous origin of the ores, opinion has been gradually crystallizing in favor of the magmatic origin, largely on account of the studies of Barlow, Coleman, and Walker, notwithstanding several vigorous protests, notably that of Dickson. However, the latest paper [47] on the subject reopens the whole question.

Our work substantiates many of the findings of Dickson [48] and of Campbell and Knight [49] relative to the Sudbury ores. We find, as they did, that the sulfids were formed later than the silicates, and verify Dickson's conclusion that the amount of hornblende increases as the ores become richer. With the exception that magnetite is later, not earlier, than the silicates, we agree with Campbell and Knight as to the order of formation of the ore-minerals.

We disagree with Dickson as to the hydrothermal ("secondary aqueous") origin of the ores.

We agree, on the other hand, with the supporters of the magmatic hypothesis that the ores were formed within the magmatic period. They were, however, not formed at an early stage and not by the sinking of the sulfid constituents.

[47] Knight, C. W.—Loc. cit.

[48] Dickson, C. W.—The ore-deposits at Sudbury, Ontario. Trans. Am. Inst. Min. Eng., 34, 3-67 (1903).

[49] Campbell, W., and Knight, C. W.—On the microstructure of nickeliferous pyrrhotite. Econ. Geol., 2, 350-356 (1907).

In fine, our work reconciles the almost diametrically opposite views of these two groups of investigators. Altho the ores are believed to be magmatic, they have been formed at the end of the magmatic period by the replacement of the silicates.

BIBLIOGRAPHY OF THE SUDBURY ORE DEPOSITS

Adams, F. D.—On the igneous origin of certain ore deposits. Can. Min. Rev., Feb. 1894.

Argall, P.—Nickel, the occurrence, geological distribution, and genesis of its ore deposits. Proc. Col. Sci. Soc., 4, 395-421 (1891-3).

Bain, J. W.—A sketch of the nickel industry. Ont. Bureau of Mines. 9th Ann. Rept., 213-224 (1900).

Barlow, A. E.—On the nickel and copper deposits of Sudbury, Ontario. Ottawa Nat., 5, 51-71 (1891).

——The Sudbury district, Ontario. Can. Geol. Surv., Summary Rept., 1901, pp. 141-145 (1902). *Idem*, 1902, pp. 252-267.

——Report on the origin, geological relations, and composition of the nickel and copper deposits of the Sudbury district, Ontario, Canada. Can. Geol. Surv. Ann. Rept. 14, pt. H, 236 pp. (1904).

——On the origin and relation of the nickel and copper deposits of Sudbury, Ontario, Canada. Econ. Geol., 1, 454-466, 545-553 (1906).

Bell, Robert.—The nickel and copper deposits of the Sudbury district, Canada. Geol. Soc. Can., 2, 125-137 (1891).

——Report on the Sudbury mining district. (Appendix by George H. Williams.) Can. Geol. Surv. Reports, Report F (1893).

Browne, D. H.—The composition of nickeliferous pyrrhotite. Eng. and Min. Jour., 56, 565-566 (1893).

——Segregation in ores and mattes. Can. Rec. Sci., 7, 176-190 (1896).

——Notes on the origin of the Sudbury ores. Econ. Geol., 1, 467-475 (1906).

Bush, E. R.—The Sudbury nickel region. Eng. and Min. Jour., 57, 245-246 (1894).

Campbell, Wm., and Knight, C. W.—On the microstructure of nickeliferous pyrrhotite. Econ. Geol., 2, 350-366 (1907).

Coleman, A. P.—The Sudbury nickel deposits. Ont. Bur. Mines, Rept. for 1903, pp. 235-299.

——The northern nickel range. Ont. Bur. Mines, Rept. 1904, pt. 1, 192-222.

——The Sudbury nickel eruptive. Geol. Soc. Am. Bull., 15, 551 (1904).

——Geology of the Sudbury district. Eng. and Min. Jour., 79, 189-190 (1905).

——The Sudbury nickel field. Ont. Bur. Mines, Rept. 1905, 14, pt. 3, 188 pp.

——Magmatic segregation of sulphide ores. Abstract, British Assn. Adv. Sci., Rept. 75th Meeting, 400 (1906).

——The Sudbury laccolithic sheet. Jour. Geol., 15, 759-782 (1907).

——Die Sudbury Nickelerze. Zeit. f. prakt. Geol., 15, 221 (1907).

——The Sudbury nickel ores. Geol. Mag., 5, 18-19 (1908).

——Copper and nickel deposits of Canada. Abstract, British Assn. Adv. Sci., Rept. 79th Meeting, 479-480 (1910).

——The nickel industry, with special reference to the Sudbury region, Ontario. Can. Dept. Mines, Mines Branch. Publ. 170, 260 pp. (1912).

——The Sudbury area. 12th Inter. Geol. Cong. Guide-Book no. 7, 48 pp. (1913).

——A classification of the Sudbury deposits. Trans. Can. Min. Inst., 16, 283-288 (1913).

——Discussion of (Petrographical notes on the Sudbury nickel deposits—Howe). Econ. Geol., 10, 390-393 (1915).

Dickson, C. W.—Note on the condition of nickel in nickeliferous pyrrhotite from Sudbury. Eng. and Min. Jour., 73, 660 (1902).

——Note on the condition of platinum in the nickel-copper ores from Sudbury. Am. Jour. Sci., (4) 15, 137-139 (1903).

——The ore deposits of Sudbury, Ontario. Trans. Am. Inst. Min. Eng., 34, 3-67 (1909).

Garnier, J.—Mines de Nickel, Cuivre, et Platine du District de Sudbury, Canada. Mem. Soc. des Ing. Civiles, Paris, 1891.

Goodwin, W. L.—Am. Jour. Sci., (3) 47, 312-314 (1894).

Gregory, J. W.—Origin of the Sudbury nickel ores. Geol. Mag., (new series) 5, 139-140 (1908).

Hixon, H. W.—Geology of the Sudbury district. Eng. and Min. Jour., 79, 334-335 (1905).

——The ore deposits and geology of the Sudbury district. Jour. Can. Min. Inst., 9, 223-235 (1906).

——The Sudbury nickel region. Eng. and Min. Jour., 82, 313-314 (1906).

Hore, R. E.—Origin of the Sudbury nickel and copper deposits. Min. and Eng. World, 36, 1345-1349 (1912).

——Magmatic origin of Sudbury nickel-copper deposits. Mich. Geol. and Biol. Surv., Pub. 16, pp. 11-37 (1914).

Howe, E.—Petrographical notes on the Sudbury nickel deposits. Econ. Geol., 9, 505-522 (1914).

Kemp, J. F.—An outline of views held today on the origin of ores. Mineral Industry, 4, 755 (1895).

Knight, C. W.—Origin of Sudbury nickel-copper deposits. Eng. and Min. Jour., 101, 811-812 (1916).

Miller, W. G.—On some nickeliferous magnetites. Brit. Assn. Adv. Sci., Rept. 1897, pp. 660-661 (1898).

Miller, W. G., and Knight, C. W.—Sudbury, Cobalt, and Porcupine geology. Eng. and Min. Jour., 95, 1129-1133 (1913).

Penfield, S. L.—On pentlandite from Sudbury, Ontario, Canada, with remarks upon three supposed new species from the same region. Am. Jour. Sci., (3) 45, 493-494 (1893).

Silver, L. P.—The sulphide ore bodies of the Sudbury region. Jour. Can. Min. Inst., 5, 528-551 (1902).

St. Clair, S.—Origin of the Sudbury ore deposits. Min. and Sci. Press, 109, 243-246 (1914).

Stewart, L.—The Creighton mine of the Canadian Copper Co., Sudbury Dist., Ontario. Jour. Can. Min. Inst., 11, 567-585 (1908).

Stokes, R.—The Sudbury nickel-copper field. Min. World, 27, 637-639, 799-801 (1907).

Stutzer, O.—Die Nickelerzlagerstätten bei Sudbury im Kanada. Zeit. f. prakt. Geol., 16, 285-287 (1908).

Thomas, K.—The Sudbury nickel district, Ontario, Canada. Min. and Sci. Press, 105, 433 (1912).

——The Sudbury nickel district of Ontario. Eng. and Min. Jour., 97, 152-154 (1914).

Thompson, P.—The Sudbury nickel region. Eng. and Min. Jour., 82, 3-4 (1906).

von Foullon, H. B.—Über einige Nickelerzvorkommen. Jahr. d. k.k. geol. Reichsanstalt, 42, 223-310. Vienna (1892).

Walker, T. L.—Notes on nickeliferous pyrite from the Murray mine, Sudbury, Ont. Am. Jour. Sci., (3) 47, 312-314 (1899).

——Geological and petrographical studies of the Sudbury nickel district. Quar. Jour. Geol. Soc., 53, 40-66 (1897).

——Certain mineral occurrences in the Worthington mine, Sudbury, Ontario, and their significance. Econ. Geol., 10, 536-543 (1915).

Williams, G. H.—Notes on the microscopical character of rocks from the Sudbury mining district. Can. Geol. Surv. Reports, (new series) 5, pt. I, Report F (1893).

THE ALEXO MINE, ONTARIO, CANADA

Geology

This occurrence of nickeliferous pyrrhotite, twenty miles southeast of Porcupine, Ontario, has been mentioned briefly by Coleman,[50] and described in detail by Uglow.[51] The geological relations are not disclosed on account of lack of exposure, and the mineralogical relations are complicated by an intense serpentinization. The ore occurs as a lens in serpentinized rock at the contact with rhyolite. In cases of this kind microscopic work is especially valuable. The unaltered rock is considered by Uglow to have been a peridotite of the wehrlite or harzburgite variety. The only exposure is in the immediate vicinity of the ore body, and developments are not sufficient to show the structural relations.

Uglow's Conclusions

Uglow examined the ore in both thin and polished sections, has shown the relations in photographs, and has brought out in a convincing way the phenomena of replacement as shown by his microscopic study. He states that the ore (pyrrhotite, pentlandite, chalcopyrite) "eats its way through the matrix of the serpentine" . . . "replaces it, forming a network of ore" . . . "extends between the crystals into fractures and cracks in the latter" . . . that "ore replaces part or all of an olivine crystal" and produces pseudomorphs—"magnetite pseudomorphs

[50] Coleman, A. P.—The Alexo nickel deposit. Jour. Geol., 5, 373-376 (1910).
[51] Uglow, W. L.—A new nickel occurrence in northern Ontario. Jour. Can. Min. Inst., 14, 657-677 (1911).

which have resulted from the alteration of the olivine, have become partly or wholly replaced by pyrrhotite," etc.

He finds that the order of the formation of the ore-minerals is (1) magnetite, (2) pyrrhotite, (3) pentlandite and chalcopyrite; and re-marks: "It is difficult to conceive of the sulphides as differentiations from a molten magma when they are as a matter of fact deposited one after the other, the younger ones occurring as vein-like masses in the older."

His work is incomplete in that he does not investigate what por-tion of the phenomena of replacement and ore migration is to be con-nected with serpentinization, and what portion with the first deposition of the ores prior to serpentinization. Lindgren [52] states: "Here, as in so many other cases, secondary changes seem to have been confused with primary deposition."

We have, therefore, investigated in detail the relative amount of transfer and migration that is connected with serpentinization, and have been able to show that the phenomena of replacement mentioned above antedate serpentinization, and are connected with the original deposition of the ores.

MICROSCOPIC DESCRIPTION

Polished sections of ore of the Alexo mine kindly furnished by Mr. F. H. Mason of the Canadian Government Exhibition Commission, show pyrrhotite with minor amounts of pentlandite and chalcopyrite in a gangue of serpentine. The serpentine is the result of alteration of oli-vine. The general relations of the minerals are shown in fig. 39. Resid-ual cores, apparently of olivine, prove to be serpentine upon examination of the thin sections. See also fig. 40, which is an enlargement of a por-tion of fig. 39. The section shown in figs. 39 and 40 was polished to bring out the silicates. On the other hand, the surface shown in figs. 41-44 was polished especially for the sulfids. Pentlandite occurs in fair amounts rather evenly distributed through the ore, with a tendency to alignment along crystallographic directions of the pyrrhotite which it replaces. The other sulfid present is chalcopyrite, which occurs in two, or possibly three, generations. In fig. 41 there is a little magmatic chalcopyrite and pentlandite. The serpentinization has been accompanied by a migration of nickel to form a second generation of pentlandite along veinlets (fig. 42), and of copper to form a second generation of chalcopyrite (fig. 44). The minute euhedral crystals of chalcopyrite shown in fig. 44 may pos-

[52] Lindgren, W.—Mineral Deposits, 765.

sibly represent a third generation of chalcopyrite. Altho there has been extensive alteration of the silicates in the Alexo ore, microscopic study shows clearly (see especially fig. 43) that this alteration was subsequent to the main period of ore formation and that the migration of ore during serpentinization was relatively slight.

SUMMARY

Summarizing the stages in the Alexo ore deposit we have (1) the formation of pyrrhotite, pentlandite, and chalcopyrite in the order named, at the end of the magmatic period, probably by selective replacement of the ground-mass of a picrite (pseudo-porphyritic peridotite); (2) the alteration of olivine and possibly of other silicates to serpentine; (3) accompanying the serpentinization there was a slight migration of copper and nickel to form second generations of chalcopyrite and pentlandite respectively, and also the formation of magnetite in very minute crystals and veinlets.

BIBLIOGRAPHY OF THE ALEXO ORE DEPOSITS

Coleman, A. P.—The Alexo nickel deposit. Econ. Geol., 5, 373-376 (1910).
——The nickel industry with special reference to the Sudbury region, Ontario. Can. Dept. Mines, Mines Branch. Report 170, pp. 112-114 (1912).
Uglow, W. L.—The Alexo nickel deposit, Ontario. Ont. Bur. Mines. Ann. Rept. 20, pt. 2, 34-39 (1911).
——The Alexo mine. A new nickel occurrence in northern Canada. Jour. Can. Min. Inst., 14, 657-677 (1911).
——Summary report on the Sudbury nickel field. Can. Dept. Mines, Mines Branch, Summary Rept. 1911, pp. 87-89.

THE FRIDAY MINE, SAN DIEGO COUNTY, CALIFORNIA

The occurrence of nickel ores in the Friday mine, San Diego county, California, was reported by Merrill [52a] and described by Calkins,[52b] whose paper furnished the geological data summarized in the following paragraphs:

The mine is four miles from the town of Julian, which lies halfway between San Diego and the Salton Sea. It is located on the crest of the broad range which forms the northern extension of the Cordillera of Lower California.

[52a] Merrill, F. J. H.—Geology and Mineral Resources of San Diego and Imperial counties. California State Mining Bureau Report, biennial period 1913-1914, p. 40.
[52b] Calkins, F. C.—An occurrence of nickel ore in San Diego county, California. S. Bull. 640, U. S. Geol. Surv., 77-82 (1916).

The complex batholith of Lower California extends into San Diego county, and the granitic rocks of the region are probably closely related to the larger masses to the south. Metamorphosed sediments are included within the granite, and near the mine a mass of gabbro of unknown size ⋅ lies in contact with mica schist.

The ore occurs as a shoot or lens at the contact of the gabbro with the schist. The latter dips steeply southward, and the ore lies at the base of the gabbro. Considerable fracturing is reported in the vicinity of the ore body, and the ore is penetrated by a small pegmatite dike containing conspicuous crystals of tourmaline.

MICROSCOPIC DESCRIPTION

Thru the kindness of Mr. Beecher Sterne, president of the Friday Copper Mines Company, we obtained a suite of rocks and ores from the Friday mine, including specimens from the recently developed lower levels. The rock is an olivine gabbro with plagioclase (Ab_8 An_7 in one section), olivine, and both orthorhombic and monoclinic pyroxene. A pale brown hornblende occurs as rims around the border of, and as patches within, the pyroxene. The hornblende is in parallel position with the pyroxene, and is doubtless a magmatic alteration product. Small tremolite prisms replace the pyroxene and plagioclase. Another alteration product is calcite, which occurs in veinlets and occasionally in zonal crystals.

The polished sections of the massive ores consist largely of pyrrhotite with residual spots of silicates. Associated with the pyrrhotite is a considerable amount of pentlandite. Calkins reported this as polydymite, but it has the characteristic cleavage, relief, and color of pentlandite. Chalcopyrite in small amounts is also present in the sections, and is distinctly later than the pyrrhotite and pentlandite. A second generation of pentlandite is developed along cracks in the pyrrhotite. A brass-yellow mineral occurs in veinlets and reticulate masses as a replacement of pyrrhotite. This was called pyrite by Calkins, but more probably it is marcasite. The marcasite gives the impression of being a very late mineral. It is usually extensively developed along calcite veinlets which occur as a net of intersecting stringers cutting all the other minerals. The polished sections also show that tremolite is later than the sulfids.

SUMMARY

In the Friday deposit we have the following sequence of events: (1) The crystallization of olivine, pyroxenes, and plagioclase. (2) A slight development of hornblende by magmatic alteration. (3) The formation of pyrrhotite, pentlandite, and chalcoypite in the order named by the replacement of the above mentioned silicates. (4) The development of tremolite as a hydrothermal mineral and the development of pentlandite of a second generation in cracks in pyrrhotite. (5) The extensive development of calcite and marcasite veinlets.

With the exception of the marcasite and calcite this deposit is similar in mineral composition and paragenesis to magmatic deposits of this type.

THE GOLDEN CURRY MINE, ELKHORN, MONTANA[53]

GEOLOGY

An interesting deposit of magmatic gold- and copper-bearing pyrrhotite (non-nickeliferous, however) occurs in a marginal "basic" segregation of the Boulder quartz monzonite batholith. The deposit thus differs from the normal type of magmatic pyrrhotite deposits in respect to the geological relations and the character of the mother rock. The mineral composition, the inclusion of the ore in a subsilicic rock, and the relatively unaltered condition of the ore and country rock, give this deposit the characteristics of a magmatic ore. It is possibly the unmetamorphosed equivalent of the lenses of pyrrhotite in "basic" layers of gneiss, and is an indication that certain of these puzzling occurrences are altered magmatic segregations.

The Boulder batholith is well known to students of ore deposits because it was accompanied and followed by extensive mineralization, and especially because it encloses the enormous copper deposits at Butte.

The Golden Curry mine at Elkhorn is in a region of intense mineralization,[54] which occurred in and near the roof of the batholith. In this property there are two types of ore deposits: (1) a contact deposit between quartz monzonite and limestone, consisting of magnetite and some chalcopyrite accompanied by garnet; (2) the magmatic pyrrhotite deposit. The latter occurs 250 feet from the contact mentioned above, as the sulfid rich portion of a lens of fresh monoclinic pyroxene, pyrrhotite, and chalcopyrite. This is surrounded by a border zone, which grades into the normal quartz monzonite by the addition of, first, plagioclase, then hornblende, quartz, and the accessory minerals of the quartz monzonite, and finally biotite and orthoclase. According to

[53] Knopf, A.—A magmatic sulphide ore-body at Elkhorn, Mont. Econ. Geol. 8, 323-336 (1913).

[54] Knopf, A.—Ore deposits of the Helena mining region, Montana. Bull. 527, U. S. Geol. Surv. (1913).

Knopf, the deposit is characterized by the absence of pneumatolytic or hydrothermal alteration products. In the vicinity, however, intense pneumatolytic action is shown in the Queen mine, the ore of which is argentiferous galena, accompanied by quartz and tourmaline. Strong mineralization of a later cooler phase is also represented in the highly productive Elkhorn mine. The ore is a replacement deposit in dolomite, consisting of argentiferous galena unaccompanied by metasomatic gangue minerals.

The Boulder batholith develops in places a border zone distinctly more basic than the normal quartz monzonite. The ore deposits, however, with the exception of the Golden Curry mine, are connected with the "acid" differentiation products, as emphasized by Billingsley.[55] Aplite masses, aplite and pegmatite dikes, often accompanied by tourmaline and sulfids, are well developed in and near the roof of the batholith. Contact deposits are common, with axinite, tourmaline, etc., in addition to the usual minerals. Tourmaline-copper and tourmaline-lead-silver ores are the high-temperature phases brought about by the mineralizing action of the intrusive. It is therefore of considerable interest to know whether the "basic" segregation of the Golden Curry mine was formed at an early or a late stage. Was there a gap between the magmatic deposits and the later high-temperature phase so extensively developed in this region?

MICROSCOPIC DESCRIPTION

Mr. Adolph Knopf of the United States Geological Survey has kindly furnished us with thin sections of the pyrrhotite-augite ore from the Golden Curry mine. The rock is practically a pyroxenite, consisting mainly of monoclinic pyroxene (augite) and pyrrhotite, with minor amounts of hornblende, tremolite, and chalcopyrite.

The replacement of pyroxene by the pyrrhotite and chalcopyrite is well shown in fig. 45. The ore-minerals not only surround the pyroxene, but occasionally cut straight across a pyroxene crystal, as shown a little to the left and below the center of the photograph. The replacement of the pyroxene by the sulfids is also proved by fig. 47, for the small pyroxene crystal near the center is cut in two by a veinlet. The pyroxene has been altered to green hornblende (not uralite) in occasional patches (the darker gray spots in figs. 45 and 47). Definite proof that the sulfids are later than the hornblende and replace it along cleavage lines is shown in the lower portion of fig. 46. Tremolite is

[55] Billingsley, P.—The Boulder batholith of Montana. Tras. Am. Inst. Min. Eng., 51, 31-56 (1915).

formed at a still later stage, for as shown in fig. 47, it projects from the end of a hornblende crystal into the sulfids.

We agree with Knopf that the pyrrhotite-chalcopyrite ore of the Golden Curry mine is a magmatic deposit showing very little post-mineral alteration, but we believe that the sulfids were formed at a late magmatic stage, and after the partial magmatic alteration of pyroxene to hornblende.

PROSPECT HILL, LITCHFIELD, CONNECTICUT

An interesting series of sulfid-bearing igneous rocks varying from gabbro and norite to peridotite and pyroxenite was described by Hobbs [56] from this locality. Extreme magmatic differentiation has produced a great variety of rock types within a small area.

Recently Howe [57] has studied the relation of the sulfids to the silicates in these rocks. He identifies pyrrhotite, pentlandite, and chalcopyrite, which were formed in the order named. A prominent feature is the presence of hornblende and biotite as magmatic alteration products of pyroxene, but no mention is made of any post-magmatic alteration. Howe concludes that the ore-minerals mentioned are magmatic sulfids which separated for the most part at an early stage in the cooling of the magma, but remained liquid until the silicates had crystallized. He believes, however, that the sulfid veinlets in the silicates were formed by the replacement of the latter.

Dr. Howe has kindly sent us some specimens of these interesting sulfid-bearing rocks, and we have corroborated nearly all of his findings. The Litchfield sulfids are magmatic sulfids, if such exist. There is no evidence of any rearrangement of the ore-minerals. The veinlets do not belong to a later generation, and all the ore-minerals are late magmatic. Altho the rocks are exceptionally fresh, perhaps more so than the average unmineralized igneous rock, we find some serpentinization of the olivine. This is especially well shown in the polished sections. [58]

We find that there has been slight post-magmatic alteration of the silicates, but conclude with Howe [59] "that the relations of the sulphides to one another were . . . determined during the magmatic period."

---

[56] Festschrift Harry Rosenbusch, 25-48. Stuttgart, 1906.

[57] Howe, E.—Sulphide-bearing rocks from Litchfield, Conn. Econ. Geol., 10, 330-347 (1915).

[58] The alteration products are very apparent in well polished sections, tho of course the identity of the minerals must be determined in thin sections or crushed fragments.

[59] Loc. cit., 339.

KNOX COUNTY, MAINE

Bastin [60] has described in detail a pyrrhotitic peridotite from this locality. The peridotite is thought to be a differentiate of granite. Chalcopyrite and pyrite are associated with the pyrrhotite, but the relation of the pyrite to the other ore-minerals is not discussed. Hornblende is locally important, and occurs both as crystals and as rims between feldspar and olivine. The secondary alteration (serpentinization) is later than the ore. His photographs show pyrrhotite surrounding the olivine and penetrating it in embayments.

This is evidently a typical magmatic deposit, with the ore later than the primary silicates.

MOUNTAIN, WISCONSIN[61]

This deposit occurs in a basic dike 60 to 200 feet wide. The ore is massive pyrrhotite which "merges into gabbro on indefinite lines." The author does not give the results of a careful field study, and no microscopic investigation of the rock or ore is recorded.

OTHER PYRRHOTITE DEPOSITS IN THE UNITED STATES

In addition to its occurrence in the magmatic ores, pyrrhotite is also found in other high-temperature deposits. Among these are the lenticular ore-bodies found in gneisses and schists. Those in the basic portion of the gneiss may be magmatic in origin, and those in the acid layers may be related to "acid" or pegmatitic extracts. Altho some of the pyrrhotite deposits occurring in metamorphic rocks, for example that of the Gap mine, Lancaster, Pennsylvania,[62] have been assigned to the magmatic group, most of them have been so modified by intense metamorphism that they no longer show the characteristics of typical magmatic deposits.

INSIZWA RANGE, EAST GRIQUALAND, SOUTH AFRICA

Our knowledge of the deposits near the town of Mount Ayliff, in Cape Colony, is due chiefly to the excellent, tho brief, report of Du

---

[60] Bastin, E. S.—A pyrrhotitic peridotite from Knox county, Maine. Jour. Geol., 16, 124-138 (1908).

[61] Bagg, R. M.—The discovery of pyrrhotite, with a discussion of its probable origin by magmatic differentiation. Econ. Geol., 8, 369-373 (1913).

[62] Kemp, J. F.—The nickel mine at Lancaster Gap, Pennsylvania, and the pyrrhotite deposits at Anthony's Nose, on the Hudson. Trans. Am. Inst. Min. Eng., 24, 620-633 (1894).

Toit.[63] He has discussed in a satisfactory manner the general geology and the petrography of the region. His microscopic description of the ores is brief, and it is chiefly in this respect that further information is needed. Unfortunately we have not been able to procure specimens from these deposits.

The deposits show certain similarities to those of the Sudbury district. The ore-minerals are the same, and they occur at the base of a large "basic" sill. They differ from the Sudbury deposits in certain respects, however, and features that are hidden at Sudbury are exposed by the erosion of the mountain mass of the Insizwa range. Du Toit's results are summarized in the following paragraphs.

Interest has been aroused in these deposits, first, by the discovery of copper, then nickel, and recently of platinum. The ores consist of pyrrhotite, pentlandite, chalcopyrite, with some pyrite, niccolite, and a platinum mineral, probably sperrylite. The ores occur at the base of the largest of a series of sills of olivine gabbro and olivine norite. The country rock, therefore, is more "basic" than the Sudbury "nickel intrusive." At this locality an extensive series or group of sills occurs in the Beaufort shales and sandstones of the Karoo system. The sills are flat-topped, but the under surface is funnel-shaped, narrowing down into the feeding dikes. Strange to say, the intruded shales are not tilted but are "undisturbed and almost absolutely flat, bounded by an under regularly curved surface of fractures."

The ores are intimately mixed with the silicates and "intrude" the country rock to some extent. The silicates of the sill show no alteration where they adjoin the ores. In addition to the ores, a granitic extract has been segregated along the base of the sill. This occurs as a network of dikes and dikelets, from a fraction of an inch to a foot in thickness. Intense contact action has affected the sedimentary rocks, and seems to be closely related to the dikes. The product of the contact metamorphism is a hornfels, described as a quartz-cordierite-feldspar rock with abundant biotite and enstatite in places. In the calcareous layers epidote, zoisite, diopside, enstatite, wollastonite, and garnet are developed.

Du Toit's figure[64] shows relations similar to those we have found in the magmatic ores from other localities, and which we interpret as indicating a replacement of the gangue by the ore-minerals. Du Toit

[63] Du Toit, A. L.—Report on the copper-nickel deposits of the Insizwa, Mount Ayliff, East Griqualand, Cape of Good Hope. Dept. Mines, 15th Ann. Rept. of the Geol. Comm., 111-142 (1910).   [64] Loc. cit., fig. 4, p. 139.

makes the order of formation of the sulfids the reverse of that determined by Campbell and Knight for the Sudbury deposits. An examination of Du Toit's sketch [65] shows the same marginal relations of chalcopyrite to the other sulfids as found in Sudbury, and we venture to suggest that further microscopic study of the ore will show that the order of the formation of the sulfids is the same at Mount Ayliff as in all other known deposits of this type.

Summarizing, we find the following suite of events connected with intrusion and ore formation at Mount Ayliff:

(1) The intrusion of a complex family of "basic" sills connected by cross-cutting dikes.

(2) The crystallization of the silicate minerals of the gabbro and norite.

(3) The development of the sulfid ores in certain localities in the basal portion of the largest of the sills by replacement of the silicates, but unaccompanied by the development of secondary silicates.

(4) The squeezing out of a granitic extract, forming a complex set of small dikes at the base of the sill, and the development of intense contact metamorphism due to gases accompanying the "acid extract." Ore formation also occurred in this stage, as ore-minerals are found in the pegmatite and granitic dikes.

NORWAY

GEOLOGY

In Norway, Nature has constructed a series of small-scale models to illustrate the type of nickel, copper, and iron sulfid deposits of magmatic origin. These instructive examples have been described in detail by Vogt,[66] whose classic work has established the existence and characteristics of this type of ore deposits. In as much as a part of the microscopic data admit of explanation according to Vogt's hypothesis of an intrusive sulfid magma, we discuss in some detail his conclusions as to the origin of these ores.

In many of the Norwegian occurrences neither the ore nor the norite has suffered metamorphism, and the geological relations are simple, so that it seems possible to select the phenomena that are characteristic of this type of deposits. It is essential for an understanding of the problems that this be done, for in many of the deposits of nickeliferous pyrrhotite elsewhere the data are either wanting or not easily

[65] Loc. cit., fig. 3, p. 137.   [66] Cited, p. 5.

deciphered. At Sudbury, for instance, it is difficult to compile the story of ore formation from the geologic records, overcrowded with the details of the eventful igneous and metamorphic history of which ore deposition is a mere episode. There, inclusions of mafic material may be explained either as "basic" segregations or as inclusions of "the older norite," as best suits the hypothesis favored by the individual investigator. Felsic masses and dikes are variously explained as inclusions of an older granite, as "acid" segregations of the norite, or as a distinctly younger intrusive granite.

In addition to the little altered types, some of the Norwegian deposits are modified by subsequent metamorphism, and in others the ores are accompanied by metasomatic borders of garnet (fig. 5). Hydrothermal alteration with the formation of pyrite appears to be developed where pegmatitic intrusions are numerous and active mineralizers were present (fig. 3).

These gradation types are fully as instructive as the normal occurrences, and support our contention that the typical magmatic deposits are one phase of high-temperature deposition, and changes in the character of the mineralizers result in gradations towards the other types.

In Norway there are about fifty deposits of pyrrhotitic copper-nickel ores which occur exclusively at or near the margins of small norite or

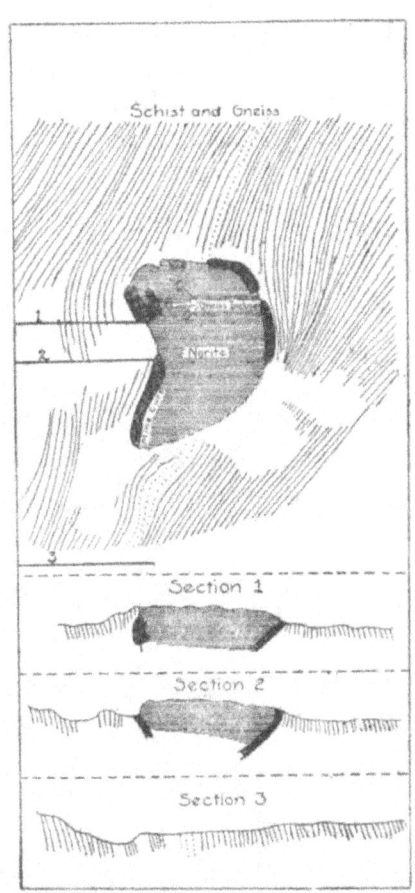

Fig. 2. Meinkjär mine, Norway. (Distance across sketch about 500 ft.) (After Vogt, Zeit. f. prakt. Geol., plate VI, fig. 3, opposite p. 133, Jahrgang 1893.)

gabbro stocks. The stocks are round or elliptical in outline, and some are funnel-shaped in cross-section. Figs. 2 and 3, taken from Vogt, illustrate typical occurrences. The individual intrusives are small, rarely exceeding a few hundred feet in diameter, and the amount of ore in each stock is roughly proportional to its size.

Petrography and Mineralogy

The ore-bearing rock may be classed broadly as norite ($\pm$ quartz, olivine, and diallage), and in a few localities it is so extensively uralitized as to be classified by some authorities as a "gabbro-diorite." Each individual stock has undergone extensive differentiation, especially in the vicinity of the ore bodies. The basic segregations are mixtures of bronzite, hornblende, olivine, and biotite. They occur in masses, smaller inclusions, and, at Romsaas, as a peculiar orbicular norite. The felsic

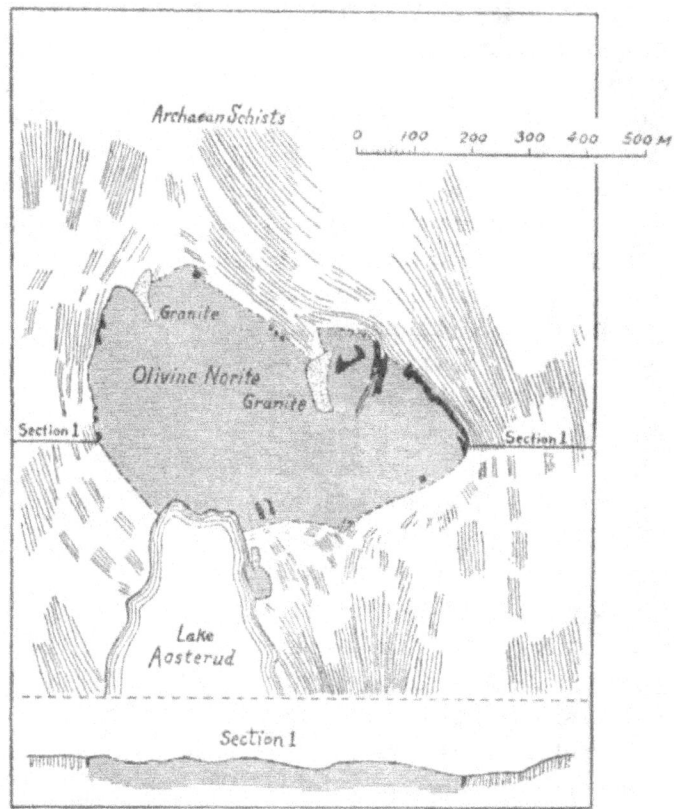

Fig. 3. Erteli mine, Ringerike, Norway. (After Vogt, Zeit. f. prakt. Geol., fig. 8, plate V, Jahrgang 1893.)

products of differentiation are as prominent as the mafic, and occur as masses of granite, streaks and dikes of pegmatitic and aplitic material, and small marginal veinlets of aplite and pegmatite. A similar differentiation appears to be a characteristic of all deposits of this class. The margins of the stocks often show severe fracturing and brecciation, and the ore may cement these igneous breccias, as at Sudbury.

According to our ideas, it would appear that (1) the "basic" portions of the norite solidified by the sinking of crystals; (2) portions of this "basic" material were caught up as inclusions in the normal norite, especially along the margins of the stocks; (3) the main mass of the rock crystallized as norite; (4) fel-
sic dikes and masses were squeezed out as marginal segregates during a period of peripheral brecciation, which facilitated the concentration and escape of mineralizers from the deeper portions of the magma; (5) ore deposition, at a late stage, was preceded, accompanied, and followed, by the "acid" secretions, as the ore is found in the felsic dikes, is later than the silicates of the same, and the main ore masses are cut by the dikes.

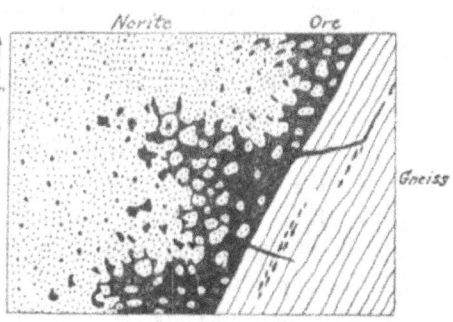

Fig. 4. Detailed cross-section of ore-body at the Meinkjär mine, Norway. (After Vogt, Zeit. f. prakt. Geol., fig. 29, p. 136, Jahrgang 1893.)

The ores occur chiefly at the margins of the stocks, and to a minor amount as segregations within the norite, and also as impregnations in the schists and gneisses in which the norite masses are enclosed. As has been mentioned, the ores are further localized where marked differen-

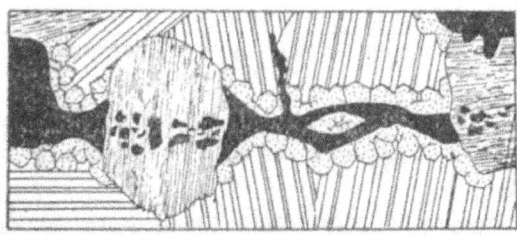

Fig. 5. Pyrrhotite gabbro from Erteli mine, Norway. Pyrrhotite veinlet cutting feldspar and diallage with rim of garnet. (x 100.) (After Vogt, loc. cit., p. 139.)

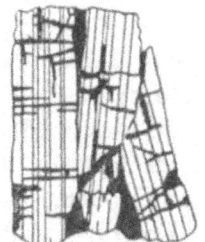

Fig. 6. Pyrrhotite veinlets cutting hornblende. Flaad mine, Evje, Norway. (x 50.) (After Vogt, loc. cit., p. 139.)

tiation has taken place. At the Flaad mine, especially, the ores are closely related to felsic differentiates.

The ores show the same group of minerals (chiefly magnetite, pyrrhotite, pentlandite, and chalcopyrite), and the same relation to each other and to the schists, as we have found in the deposits studied elsewhere. Pyrite is mentioned by Vogt as of common occurrence in these deposits. He describes it especially in the Erteli mine at Ringerike, and the Flaad mine at Evje. Ores from both these localities, furnished through the courtesy of Professor J. F. Kemp, were examined by us. The pyrite from Ringerike occurs as euhedral and irregular grains, cutting both the other ore-minerals and the silicates. Its position in the sequence of mineral formation could not be definitely determined from the material available. Pyrite, in the ore from Evje, however, clearly cuts garnet and also all the ore-minerals, including chalcopyrite, as groups of anastomising sub-parallel veinlets (fig. 48). These veinlets widen into irregular areas, and even subhedral crystals are often developed. This makes it probable that the euhedral and subhedral pyrite crystals of fig. 48, as well as the veinlets, are later than the pyrrhotite. Chlorite and calcite cut the sulfids in sharp veinlets, and occur along the pyrite veinlets. They are certainly later than all the sulfid minerals except pyrite, and may be later than it.

As already stated, pyrite is not commonly found in the magmatic ores. At Sudbury, it is found especially in the Worthington offset, connected with later veins of hydrothermal origin. In the Ringerike deposits ore has been mined from disseminations in the country rock, and the pyrite in these ores may have been developed in the schists prior to the introduction of the ores. At Evje, as stated above, the pyrite is the last sulfid mineral to form, and may have been developed during the extensive uralitization of the country rock. A complete microscopic study of all the Norwegian ores according to modern petrographic and metallographic methods is needed in order to determine the paragenesis of the pyrite.

As described and figured by Vogt (fig. 5), veinlets of sulfid ores are accompanied by garnet, showing a gradation of these deposits to other high-temperature deposits.

Contrary to the general impression, Vogt states that the ores in all cases are later than the silicates. He notes [67] the frequent occurrence of microscopic veinlets in the cracks and cleavage planes of the silicate minerals. We wish to emphasize again, that these data, checked by us, admit of only two explanations, viz., (1) that of Vogt, who postulates an intrusion of molten sulfids into the solidified rock; or (2), our hypothesis that the sulfids were formed by the replacement of silicates under the influence of mineralizers. This process, however, belongs to a late magmatic period, and not to destructive pneumatolytic or hydrothermal stages, and the deposits, therefore, may properly be designated as magmatic.

---

[67] Zeit. f. prakt. Geol., Jahrgang 1893, p. 139.

DISCUSSION OF VOGT'S CONCLUSIONS

In the light of the data summarized in the preceding pages we may now examine in detail the data listed by Vogt,[68] in his latest contribution, as favoring his hypothesis.

"As suggested more particularly by Vogt in 1893, the nickel-pyrrhotite deposits are to be regarded as magmatic segregations in gabbro or in chemically analogous dike rocks. The following points speak for such an origin:

1. The connection of the numerous deposits in different countries with occurrences of gabbro or exceptionally with equivalent dike rocks, is constant and regular.

2. The several occurrences so resemble one another not only geologically but also mineralogically that they must be of the same genesis throughout, while the unvarying character of the deposits postulates a simple process of formation.

3. Gradations between ore and gabbro through the intermediate stage of pyrrhotite-gabbro are often to be observed, and therefore the ore essentially must have been formed in a similar manner to that rock.

4. The structure of the clean or almost clean sulphide mixture with idiomorphic crystals suggests crystallization from a single magmatic solution, and not from several solutions following one another, as was evidently the case for example with the lead-silver-zinc lodes.

5. The deposits often occur in those parts of a gabbro mass which are distinguished by pronounced differentiation of the eruptive rock.

6. The deposits are sometimes crossed by dikes of basic rock, diabase, olivine-diabase, etc., which are to be regarded as later effusions of the same eruption of gabbro. The formation of the ore belongs therefore to the magmatic period of the rock in which it occurs.

7. Some of the deposits are traversed and accompanied by acid leucocratic streaks and dikes which represent the acid segregation products from the gabbro-magma, from which also it follows that the formation of the deposits took place during the magmatic period of the eruptive rock.

8. The characteristic presence of titanomagnetite allows a manner of formation analogous to that of the magmatic titaniferous-iron deposits to be postulated.

9. Pneumatolytic minerals are completely wanting.

10. The minerals usually formed in the wet way are not present as part of the primary formation."

The points in paragraphs 1 and 2 are verified by our work. The ores are so closely connected with a particular type of basic rock

---

[68] Ore Deposits: Beyschlag, Vogt, and Krusch; Translated by S. J. Truscott; pp. 286-287. We have revised the translation slightly for the sake of accuracy.

that a genetic relation cannot be doubted. The majority of the deposits are so similar that one method of ore formation alone is involved. The various complicated sets of minerals produced by hydrothermal solutions are either lacking or are developed after the formation of the ores, and hence a close relation to rock crystallization is suggested.

Paragraph 3 has little meaning. Variations in amount of replacement, from incomplete to complete, are found in most ores formed by this process.[69] The same relations obtain in magmatic ores.

In regard to paragraph 4, Vogt is in error as to the facts. Apparently he had not studied polished ore surfaces, nor has he grasped the meaning of the data set forth in the articles by Dickson [70] and by Campbell and Knight.[71] The ores of the magmatic stage are introduced into the completely crystallized rock, one after the other, in the following order: (1) magnetite and ilmenite, (2) pyrrhotite, (3) pentlandite, and (4) chalcopyrite. There is positive evidence of a certain amount of the replacement of the earlier ore-minerals by those introduced at later stages. This replacement cannot be explained by corrosion, for there are no intermediate reaction products. The replaced minerals are completely removed by the same agencies that brought in the ores. Vogt recognizes that some of the chalcopyrite is later than the other ore-minerals. He states:[72] "Die häufig beobachtete Anreicherung des Kupferkieses, namentlich in den peripheren Teilen der Nickel-Magnetkieslagerstätten muss auf eine besondere magmatische Differentiation innerhalb der Sulfidmagmen beruhen."

To apply his ideas to the facts now established, he would have to postulate a triple or quadruple differentiation within the sulfid magma.

The points in paragraphs 5, 6, and 7 are verified and emphasized by us. Magmatic differentiation is characteristic of the ore-bearing norite, and is especially marked in the vicinity of the ores. Extensive early differentiation calls for an equivalent development of the complementary later differentiation. The ores are a phase of the latter. They are extracted from the magma basin by mineralizers, and are brought to the margin of the deposit and deposited without the development of secondary silicates.

Point 8 is valid.

---

[69] Knight (loc. cit.) emphasizes this in his recent discussion of the Sudbury ores.

[70] Loc. cit.   [71] Loc. cit.

[72] Beyschlag, Krusch, Vogt.—Die Lagerstätten der nutzbaren Mineralien und Gesteine. Band 1, 284-285 (1910).

Point 9 is true in general, altho tourmaline and garnet are occasionally present and hornblendization generally precedes ore deposition. Vogt evidently assumes that high-temperature deposits formed by mineralizers must be accompanied by destructive pneumatolysis. In this regard it is interesting to note that Vogt's early studies led him to suggest a "combination magmatic segregation with pneumatolysis, according to which metalliferous vapors evolved from the magmas in one place were decomposed in another, depositing the ore in rock already consolidated."[78] It is needless to state that his early concept approached more closely to the truth than his present hypothesis.

Paragraph 10 may be more accurately stated as follows: The secondary silicates usually formed by destructive pneumatolysis and hydrothermal action are not developed in the magmatic stage of ore formation.

SUMMARY

We conclude with Vogt that the ores are so closely related to the intrusive rocks which contain them, and to the processes of rock differentiation, and differ so markedly from deposits formed by pneumatolytic and hydrothermal processes, that they should be classed as magmatic. However, they cannot be considered as segregated in the molten stage by a "liquation" process, and at a later date intruded into the silicates, for the following reasons:

1. The ore-minerals are formed in an orderly sequence, one after the other. A succession of sulfid differentiations and intrusions is beyond the realm of probability.

2. The ores replace the country rock without reaction rims, and the silicates thus replaced have been completely removed.

3. The minerals are introduced after the magmatic alteration of pyroxene to hornblende, but probably prior to the intense uralitization certain of the deposits have undergone.

4. The temperature of the late stages of differentiation in the presence of mineralizers, and the formation of pegmatite and aplite dikes, is far below the temperature of molten ores, as we know them as furnace products.

5. The phenomena of ore formation and rock replacement are similar in all respects to that of sulfid ore deposition from hydrothermal solutions, except for the comparative absence of secondary silica and silicates. This is a characteristic feature of this type of ore, and is due to the conditions of chemical equilibria under which they are formed.

---

[78] Truscott trans. Loc. cit., 289.

BIBLIOGRAPHY OF THE NORWEGIAN DEPOSITS

Vogt, J. H. L.—Bildung von Erzlagerstätten durch Differentiationsprocesse in basichen Eruptivmagmata. Zeit. f. prakt. Geol., Jahrgang 1893, pp. 4-11, 125-143, 257-284.

See also Stelzner-Bergeat, Die Erzlagerstätten, I, 48, 1904, for literature of the Norwegian occurrences.

OTHER EUROPEAN OCCURRENCES

SWEDEN

The nickel- and copper-bearing pyrrhotite deposits of Sweden are similar to the more extensive group in Norway. They appear, in general, to have suffered greater alteration than the latter. The largest deposit is at Klefva, in Småland. The ores occur in a quartz norite which has suffered intense uralitization. Some deposits at Kuso and Stattberg occur in mafic (diabase) dikes, and are somewhat similar to the occurrence at Sohland mentioned below.[74]

BADEN, HORBACH, AND TODTMOOS, GERMANY

In the southern Black Forest pyrrhotitic ores occur in small amount in basic inclusions and dikes, extensively altered to amphibolite and serpentine, which occur in granite and gneiss. They are cut by granite-aplite dikes.[75]

SOHLAND AND SWEIDERICH ON THE BOUNDARY BETWEEN SAXONY AND BOHEMIA

Nickel- and copper-bearing pyrrhotite[76] impregnates the footwall of irregular "basic" dikes which are varied in composition on account of pronounced differentiation.

The ores, according to Beck, surround and cut corroded augite, hornblende, and biotite, and shatter and impregnate the latter. Beck was

---

[74] Literature is cited, Beyschlag, Krusch, Vogt, I, p. 294, and Stelzner-Bergeat, I, p. 48.

[75] Weinschenk, E.—Die Nickelmagnetkieslagerstätten in Bezirk St. Blasien. Zeit. f. prakt. Geol., Jahrg. 1907, pp. 73-86. Other literature cited in the textbooks of Beck and Stelzner-Bergeat.

[76] Beck, R.—Die Nickelerzlagerstätten von Sohland a. d. Spree und ihre gesteine. Zeit. deutsch. geol. Ges., Jahrg. 1903, pp. 296-331. (Includes bibliography of the older literature.)

——Lehre von Erzlagerstätten, I, 81-86 (1909).

von Foullon, H. B. Über einige Nickelvorkommen. Jahrb. k.k. geol. Reichsanst. 302 pp. Wein. (1892).

unable to distinguish pentlandite from pyrrhotite, but we found no difficulty in showing its existence in polished surfaces (see figs. 50 and 51, plate XII).

Beck has proved conclusively that the ores are later than the primary silicates, and believes that they are formed by a "post-volcanic, pneumatolytic phase of rock building."[77]

We have a small suite of specimens from Sohland which include both the unaltered rock and ores. These we have examined in thin sections, and the ores also in polished sections. The unaltered rock is a hornblende-diabase (called proterobase by Beck), containing plagioclase, augite, brown hornblende, biotite, apatite, titanite, and magnetite. The only alteration products present are a little calcite, tremolite, and chlorite. The hornblende occurs in parallel position with the augite and is evidently a late magmatic mineral.

In thin sections of the ore the only original silicate mineral noted is the hornblende. The feldspars and biotite are altered almost beyond recognition. The alteration products are chlorite, sericite, and uralite. Of especial interest is the uralite, because its relation to the ore-minerals can be definitely established.[77a]

Fig. 50 shows the general relations of the minerals in the polished sections. A study of our sections together with Beck's original figures (especially his fig. 4 of plate XIII) makes it almost certain that the ore-minerals were formed later than the hornblende, but earlier than the uralite. Uralite develops on the ends of the hornblende prisms in parallel position, and occasionally completely replaces the hornblende. Frequently the uralite needles project out into the ore-minerals, as in fig. 51, and in Beck's fig. 4. This affords conclusive evidence of post-mineral alteration. For this reason we believe that the Sohland ore is magmatic in spite of the extensive alteration. The relation of the ore-minerals to the silicates was evidently established during the magmatic period, and not modified by later hydrothermal action.

---

[77] Quoted by Berg, opus cit., 108.

[77a] In the literature the term uralite has been used in two senses, (1) for hornblende rims around pyroxene, (2) for the fine fibrous aggregates of tremolite (including actinolite). The first usage we have avoided entirely, and in the few cases in which the terms uralite and uralitization are used by us, these are to be given the second meaning. The hornblende rims appear to be late magmatic and the tremolite a post-magmatic, hydrothermal mineral.

GROUP II. MAGMATIC CHALCOPYRITE-BORNITE DEPOSITS

OOKIEP, NAMAQUALAND, SOUTH AFRICA
GEOLOGY

The important mines of this region occur as great lenticular deposits in the vicinity of the town of Ookiep, Namaqualand, Cape Colony, South Africa. They lie about sixty miles east from the Atlantic coast, and about twice as far south of the Orange river.

The ore bodies occur as shoots or lenses, and again as disseminated sulfid particles, in a system of dikes and sills, which give evidence of remarkably extensive differentiation. Rogers[78] describes the "ore bearer" as including rock varieties made up almost wholly of any one of the following minerals: magnetite, quartz, plagioclase, hypersthene, hornblende, and biotite, and also intermediate varieties such as norite, mica diorite, augite diorite, and diorite. A single dike may show one or more varieties of rock, and where several kinds occur in one intrusive the contacts are sharp, suggesting differentiation before intrusion.

According to Kuntz[79] the intrusive rock occurs as dikes of great extent, or again as detached stocks. To these Rogers adds: pipes, nearly horizontal sheets, and branching bodies; and states that 344 intrusive bodies have been mapped up to date. These dikes cut the "fundamental gneiss" of South Africa, and the ore occurs as shoots, occupying a portion of the entire width of the dikes, and also penetrates the gneiss adjacent to the dikes.

The principal ore-minerals are given as magnetite, pyrrhotite, bornite, and chalcopyrite. We find that hematite is also an important constituent of the ores.

The chief mines are Ookiep, Specktakel, Nababeep, and Ookiep East, all situated near the town of Ookiep, and the Tweefontein mine is located on a dike to the north. The ore shoots are lenticular masses of high grade ore with greatest dimension in a horizontal direction. For example, the Ookiep ore-body, according to Ronaldson,[80] is 1300 feet

---

[78] Rogers, A. W.—The nature of the copper deposits of Little Namaqualand, Proceed. Geol. Soc. S. Africa, 1916, pp. xxi-xxxix.

[79] Kuntz, J.—Copper ore in southwest Africa. Trans. Geol. Soc. South Africa, 7, 70-76 (1904).

[80] Ronaldson, J. H.—Notes on the copper deposits of Little Namaqualand. Trans. Geol. Soc. S. Africa, 8, 158-167 (1905).

wide and 300 feet deep. The Tweefontein mine contains a series of three lenses, one under the other.

On account of the occurrence of the copper ore intimately mixed with apparently unaltered constituents of the dikes, a magmatic origin was early assigned to these deposits. In 1857, Wiley[81] stated that the ores were of magmatic origin, and Daintre in 1878 makes the plain statement that "the ores are of magmatic origin, occurring in unaltered feldspathic dikes from specks the size of a pin point to many tons."[82]

Microscopic Descriptions

We are indebted to Dr. A. W. Rogers, Director of the Geological Survey of the Union of South Africa, for an excellent suite of specimens of ores and accompanying rocks from the Ookiep mines. These rocks vary from an almost pure hypersthene rock (hypersthenite), through norite and mica diorite, to an almost pure plagioclase rock (anorthosite). The specimens include both lean and rich norite and mica diorite, but the anorthosite is almost free from the ore-minerals, and the hypersthenite carries a fair amount of the ore-minerals.

Ore-bearing Hypersthenite from the Nababeep Mine.—The general relations of the ore-minerals to the silicates of the hypersthenite are well shown in fig. 54. The ore-minerals occur in anhedra which are often elongate and pass into vein-like forms. It is probable that bornite has formed by the replacement of the magnetite, for they show the same general shape.

Fig. 52 represents a thin section of the hypersthenite. The transparent mineral is hypersthene. The black area is mainly magnetite, but a little bornite is also present, and in some places in the section bornite predominates over magnetite. The ore-minerals surround the hypersthene crystals, and in one place (just below the center) a veinlet of magnetite cuts across a hypersthene crystal.

Besides hypersthene the only original silicates present are biotite and plagioclase, which occur in minor amounts. The plagioclase is shown in little patches in fig. 53. The association of ores with these areas of feldspar, which considered with respect to the enclosing rock are of the nature of "acid extracts," is significant. The feldspar seems to be more readily replaced by the ore-minerals than the hypersthene. Note the small magnetite euhedra within the hypersthene.

---

[81] Wiley, A.—Report on the mineral and geological structure of South Namaqualand. Parliamentary Report, Cape Town, no. 36, p. 30 (1857).

[82] Quar. Jour. Geol. Soc., 34, 434 (1878).

The minerals of the hypersthenite are comparatively fresh, but there has been some alteration along the boundaries between the hypersthene anhedra. This alteration product of hypersthene is probably anthophyllite. It occurs in fibrous aggregates which are often intimately associated with the bornite. Fig. 55 is a photomicrograph of an area showing the relations of the anthophyllite to the bornite. A careful study of the section shows that the anthophyllite has replaced the bornite. The narrow linear areas of bornite might be considered evidence that the bornite had replaced anthophyllite; but that the reverse is true is proved by the obliquely-cutting anthophyllite needles shown at several points.

Ore-bearing Norite from the Tweefontein Mine.—Sections of ore-bearing norite from the Tweefontein mine are represented on plates XIV and XV. The silicate minerals are hypersthene and plagioclase, with subordinate biotite. The ore-minerals are magnetite, chalcopyrite, and bornite. Fig. 56 illustrates the replacement of hypersthene by ore-minerals, principally magnetite with a little bornite and chalcopyrite. At the bottom of the photograph there is a small magnetite crystal, which is anhedral along the border between the hypersthene and plagioclase and euhedral within the plagioclase crystal. This is better shown in fig. 57.

In fig. 58 the opaque mineral is largely bornite, with a little chalcopyrite and magnetite. At the bottom of the figure a hypersthene crystal is almost completely surrounded by bornite. A higher magnification of the upper left corner of this area is shown in fig. 59. The bornite has replaced the plagioclase in the direction of twinning lamellae.

Figs. 60-63 (plate XV) show the occurrence, in the Tweefontein norite, of the ore-minerals in polished sections. The general relations are shown in fig. 60. Magnetite and hematite are readily distinguished by the character of their surfaces (fig. 61). Magnetite is rough and is intergrown with ilmenite; hematite is smooth. Magnetite and hematite, as well as the sulfids, surround and replace the silicates. The replacement of biotite by the sulfids is well shown in fig. 60. Note the veinlike magnetite which cuts directly through a hypersthene crystal.

The norite specimens from the Tweefontein mine are remarkably free from alteration, as can be seen from the photomicrographs. There was, however, a little alteration along minor fractures (fig. 62) in bornite. Along one of these fractures (illustrated by fig. 63) gashes of chalcopyrite of a second generation and minute crystals of anthophyllite were developed. The high magnification of these figures shows how insignificant is the alteration.

Another specimen of norite taken from drill-cores at the Tweefontein mine constitutes a lean ore. Very small amounts of bornite and chalcopyrite are visible. There is, however, considerable magnetite. A study of the thin section shows as good evidence of the replacement of the silicates as one could wish for. A small apatite crystal has been cut squarely in two by magnetite.

Ore-bearing Mica Diorite from the Ookiep East Mine.—There remains to be described the mica diorite from the Ookiep East mine. The principal minerals are plagioclase and biotite. The alteration products include anthophyllite, clinozoisite, and chlorite. The ore-minerals are magnetite, pyrrhotite, and chalcopyrite; bornite is absent. The mica diorite is poor in ore, and, strange to say, is more altered than that containing large amounts of chalcopyrite and pyrrhotite. Fig. 65 illustrates the occurrence of the ore-minerals in the rich ore as seen in thin section. The alteration veinlet (chlorite) at the bottom of the figure is later than the sulfids. The light spots within the opaque areas are also alteration products.

A study of the polished sections with a low-power microscope indicates that the chalcopyrite is probably formed by the replacement of the pyrrhotite (see fig. 64). Another spot furnishes evidence that the pyrrhotite has in part been formed by the replacement of magnetite (see fig. 67).

Some portions of the specimens of rich ore from the Ookiep East mine contain definite sulfid veinlets. These are shown in thin section in fig. 66 and in polished section in fig. 68. These veinlets, unlike the sulfid veinlets in the Sudbury ores, have a peculiar "fuzzy" appearance which (see fig. 66) suggests that they might have been produced by rearrangement of the larger sulfid masses. The study of the thin sections under high magnification shows that this appearance is due to the presence of a secondary silicate (chlorite or anthophyllite). That the veinlets are not due to rearrangement is proved by fig. 69, which distinctly shows that the alteration product, probably chlorite formed by the alteration of anthophyllite, is later than the chalcopyrite of the veinlets. Minute specks of a light yellow mineral, probably pentlandite (fig. 71), were also found in the ore from the Ookiep East mine. It is a late, probably hydrothermal, mineral, and furnishes the only evidence of rearrangement of the sulfids in our specimens of the ores of the Ookiep East mine.

The replacement of the silicates is shown in all the photographs in plates XVI and XVII, but especially well in fig. 70. This represents a biotite-rich segregation in the diorite. The biotite, itself probably a late magmatic mineral, has been replaced along its cleavage planes by chalcopyrite. This produces a structure which can be distinguished from that developed by the replacement of the sulfids by later hydrothermal minerals such as sericite, chlorite, and tremolite. (See for examples figs. 82 and 83, plate XX.)

SUMMARY

The Ookiep ores are of importance in establishing the type of magmatic copper deposits. All who have studied these ores have classified them as magmatic ores, but in spite of this they have received scant attention by the authors of textbooks and treatises on ore deposits. The principal reason for the failure to give these deposits full standing in the magmatic class is the presence of bornite. The Ookiep ores are magmatic, if such exist. They are the least altered of any we have examined in this study. The ore-bearing (chalcopyrite-bornite) norite from the Tweefontein mine is much freer from alteration than the average igneous rock. It is true that the mica diorite and hypersthenite at Ookiep locally show some alteration, but the significant thing is that the ore-minerals in the altered and unaltered rocks show exactly the same relation to the silicate minerals.

BIBLIOGRAPHY OF THE OOKIEP DEPOSITS

Delesse, M.—Sur les mines de cuivre du Cap de Bonne Esperance. Ann. des Mines, 5me serie, 8, 186-212 (1855).

Wiley, A.—Report on the mineral and geological structure of South Namaqualand. Parliamentary Report, Cape Town, no. 36, p. 30 (1857).

Zerrener, C.—Reise des Ingenieurs A. Thiers nach den Kupfer Bergwerken Namaqualands in Sud Africa. Berg. und Hüttenm. Zeitung 1860, pp. 41-44 and 53-54.

Knopf, A.—Über die Kupfererzlagerstätten von Klein Namaqualand und Damaraland. Neues Jahrb. f. Min. Geol. u. Pal., 1861, pp. 513-550.

Schenk, A.—Die Kupfererzlagerstätten von Ookiep in Klein Namaqualand. Zeit. de deutsch. geol. Ges. Verhand. der Ges., 53, 64-65 (1902).

Kuntz, J.—Copper ore in Southwest Africa. Trans. Geol. Soc. South Africa, 7, 70-76 (1904).

——Kupfererzvorkommen in Südwestafrika. Zeit. f. prakt. Geol., 12, 199-202 (1904).

Ronaldson, J. H.—Notes on the copper deposits of Little Namaqualand. Trans. Geol. Soc. South Africa, 8, 158-167 (1905).

Stutzer, O.—Magmatische Ausscheidungen von Bornit. Zeit. f. prakt. Geol., 15, 371 (1907).

Rogers, A. W.—The nature of the copper deposits of Little Namaqualand. Proceed. Geol. Soc. S. Africa, 1916, pp. xxi-xxxiv.

ENGELS MINE, PLUMAS COUNTY, CALIFORNIA

GEOLOGY

A rather unique copper deposit, which we believe to be magmatic, is now being mined by the Engels Copper Mining Company in the northern part of Plumas county, California. This deposit has been studied and described by Turner and Rogers.[88]

The ore occurs in a remarkably fresh norite-diorite occurring at the extreme northern end of the great Sierra Nevada batholith of granodiorite. There is no evidence of dynamic metamorphism and none of contact metamorphism, for the older rocks into which the norite-diorite is intrusive are nearly five miles distant. The main ore body is a tabular ore shoot which has a nearly vertical attitude. It is difficult to see how this ore body could possibly be explained by gravitative adjustment due to the sinking of sulfids in the molten magma. Narrow pegmatite dikes are occasionally found cutting the norite-diorite.

MICROSCOPIC DESCRIPTIONS

The predominant igneous rock at the Engels mine is a norite-diorite with about 46 per cent of silica. The principal minerals are plagioclase (andesine-labradorite, $Ab_1 An_1$), hypersthene, diopside, hornblende, and biotite. The norite-diorite varies from a rather fine-grained hypersthene-plagioclase rock to a coarse-grained hornblende-plagioclase rock. One of the typical fine-grained rocks (from no. 2 level) is illustrated in fig. 72. The opaque minerals here are magnetite and hematite. They surround and replace the silicates, especially hypersthene and biotite. Evidence of replacement is shown at many spots in the photomicrograph. A magnified view of one of these is shown in fig. 74. The magnetite and hematite have surrounded hypersthene and plagioclase, and have penetrated and replaced a biotite crystal. It will be noticed in fig. 72 that the magnetite occurs in both euhedral and anhedral crystals. The euhedral crystals occur within the silicate minerals, and the anhedral crystals mainly along the boundaries of adjacent silicate anhedra. This is also well shown in fig. 73. In this figure there is a magnetite crystal which shows anhedral development on the boundary between the hypersthene and plagioclase, and euhedral development within the hypersthene. This is an argument in favor of the late magmatic origin of

[88] A geologic and microscopic study of a magmatic copper sulphide deposit in Plumas county, California, and its modification by ascending secondary enrichment. Econ. Geol., 9, 359-391 (1914).

euhedral magnetite, as is also the fact that all gradations between an-hedral and euhedral forms occur. This practically proves that the euhe-dral, as well as the anhedral, magnetite is formed by the replacement of the silicates at a late stage. This rock is practically free from alteration.

The coarse-grained norite-diorite usually contains hornblende in-stead of hypersthene. The hornblende in these rocks has probably been formed from pyroxene. Evidence of this late magmatic alteration is furnished in fig. 75, where residual cores of diopside and hypersthene occur within a hornblende crystal.

In plate XIX are shown photomicrographs of a specimen somewhat similar to the rock just described, except that it contains alteration pro-ducts. The ferro-magnesian minerals are hypersthene, diopside, and hornblende. The latter occurs as rims around the diopside (fig. 76).

The ore-minerals in this specimen are magnetite, hematite, chal-copyrite, and bornite. The magnetite contains regularly arranged ilmen-ite plates. The hypersthene is extensively altered to a mineral with indefinite aggregate structure. This contains tremolite and proba-bly talc, and it is possible that tremolite is an intermediate product of the alteration of hypersthene to talc. Chlorite, another alteration prod-uct, occurs in veinlets which definitely cut the ore-minerals as shown in fig. 77.

Covellite and chalcopyrite of the second generation have been formed in occasional spots at the expense of the bornite (fig. 81). These are the result of the rearrangement brought about subsequent to the mag-matic stage.

Some of the ore at the Engels mine occurs in a rather fine-grained felsic rock containing plagioclase, orthoclase (or microcline), quartz, and biotite, with minor accessories. For convenience this rock is called grano-diorite. In one specimen of this type apatite, titanite, epidote, calcite, and analcite were found. These are not alteration products of any minerals present in the rock, and are considered to be of late mag-matic origin and formed by mineralizers. This rock is remarkably fresh; the only hydrothermal products present are a little chlorite formed from biotite and a little sericite from feldspar.

Another type of the fine-grained grano-diorite is exceptional in that it contains tourmaline. This specimen, which is figured in plate XIX (figs. 78 and 79), contains a good deal of bornite and chalcocite. The bornite occurs in irregular anhedra which surround and replace the silicates. The tourmaline has also been replaced by bornite (fig. 78). This specimen contains considerable chlorite, which is evidently pseudo-morphous after biotite.

The igneous rocks which constitute the ores at the Engels mine contain varying amounts of alteration products, which are largely sericite, chlorite, tremolite, and talc. Some specimens, such as those represented by fig. 72, are practically free from alteration, while others are considerably altered. The hypersthene is frequently altered to a gray indefinite substance with aggregate structure (fig. 77). This is probably talc. It is probable that tremolite is an alteration product, intermediate in point of time between hypersthene and talc.

Sericite is present in certain specimens in fair amounts. It occurs as a replacement of feldspar, and also as a replacement of the sulfids. This is shown in fig. 78. The sharp lath-shaped crystals show well in contrast with the black bornite (and chalcocite), but exactly the same kind of crystals appears in the feldspars.

The chlorite occurs in definite veinlets cutting the ore-minerals (fig. 77), and also in lath-shaped sections resembling sericite and these also cut the ore-minerals (fig. 82).

The rearrangement of the ore-minerals in magmatic sulfid deposits is insisted upon by several writers to explain certain features. In the Engels ore the rearrangements brought about by lowering the temperature are well illustrated. The magnetite and hematite suffer no changes, but the bornite is often broken down into chalcocite, covellite, and chalcopyrite of the second generation. Examples of these are shown in plate XX. The so-called graphic intergrowth of bornite and chalcocite (fig. 80), as one of us suggests in a recent paper,[84] is a local, very irregular, replacement of bornite by chalcocite. The replacement of bornite by chalcocite also takes place in veinlets (fig. 83), and occasionally along crystallographic directions; but the best example of crystallographic influence during replacement is the break-down of the bornite to chalcopyrite of the second generation (fig. 81). Covellite also replaces bornite in irregular splotches (fig. 81), or along crystallographic directions (faintly shown in fig. 82).

It seems probable, as one of us [85] has stated, that a first generation of chalcocite (fig. 82) was formed before the sericite and chlorite, and a second generation (fig. 83) after sericite and chlorite. The former is hypogene,[86] the latter supergene.

[84] Rogers, A. F.—The so-called graphic intergrowth of bornite and chalcocite. Econ. Geol., 11, 582-593 (1916).

[85] Rogers, A. F.—Sericite a low-temperature hydrothermal mineral. Econ. Geol., 11, 118-150 (1916).

[86] Ransome, F. L.—Copper deposits near Superior, Arizona. Bull. 540, U. S. Geol. Surv., 152 (1912).

SUMMARY

The Engels mine, which is rapidly becoming one of the important copper mines of California, is unique in that magmatic bornite is the principal ore-mineral. The ore occurs as a vertical ore shoot, and in a massive norite-diorite, in such relation to the enclosing rock that only a magmatic origin seems probable. Bornite and associated chalcopyrite are formed by mineralizers at a late magmatic stage, and not by hydrothermal solutions. This is proved by the comparative freedom from alteration in many of the ores and associated rocks. The silicification accompanying hydrothermal deposits generally is practically absent. Locally there may be sericitization and chloritization, but the bornite and chalcopyrite are no more abundant in the ores thus affected than in the unaltered ores. The secondary copper minerals, covellite, chalcocite, and chalcopyrite of the second generation, have developed at a later stage, in part by hypogene and in part by supergene solutions.

BIBLIOGRAPHY OF THE ENGELS ORE DEPOSIT

Aubury, L. E.—The copper resources of California. Bull. no. 50, California State Mining Bureau, pp. 185-186 (1908).

Turner, H. W., and Rogers, A. F.—A geologic and microscopic study of a magmatic copper sulphide deposit in Plumas county, California, and its modification by ascending secondary enrichment. Econ. Geol., 9, 359-391 (1914).

Read, T. T.—The Engels mine and mill. Min. and Sci. Press, 111, 167-171 (1915).

REMARKS ON CERTAIN OTHER DEPOSITS THAT HAVE BEEN CLASSIFIED AS MAGMATIC

PYRITIC DEPOSITS

Beyschlag, Krusch, and Vogt [87] classify certain pyrite ores as magmatic ores under the heading "Die intrusiven Kieslagerstätten." Among these they place the Rio Tinto, Bodenmais, Sain Bel and Chessy, Agordo, and numerous Norwegian deposits. It is certain that some of these pyritic ores are hydrothermal in origin. The Rio Tinto deposits, for example, are due to metasomatic replacement of crushed and sheared zones by hydrothermal solutions, as was definitely proved by Finlayson.[88] The Rio Tinto and other pyritic deposits, such as the Rammelsberg, are placed by Lindgren [89] in his division "Deposits formed at Intermediate Depths." Lindgren, however, recognizes that some of the pyritic deposits may be due to the injection of molten sulfids.

The so-called intrusive or injected pyritic ores occur for the most part in gneisses. Igneous rocks are often closely associated, tho not always. The igneous rocks, however, are altered rocks, such as saussurite-gabbro. The ore-minerals are associated with typical metamorphic minerals such as cordierite, andalusite, anthophyllite, and chloritoid.

As these ores occur in areas of regional metamorphism, and are usually found in gneisses, schists, and not often in the igneous rocks themselves, it is difficult to see why they are placed with the magmatic ores. It seems far more reasonable to classify them as metamorphic ores, as Berg [90] does. We have not had an opportunity of examining many of these ores, but certain facts derived from our study of the undoubted magmatic ores leads to the conclusion that these are not of this type, or at least that the burden of proof rests upon any one who classifies them as such.

In the first place, we seriously doubt whether pyrite is a characteristic magmatic mineral. In our present study we have not found pyrite in any of the typical disseminated magmatic ores, and our speci-

---

[87] Loc. cit., 1, 298 (1910).

Ore Deposits. English translation by Truscott, 1, 301 (1914).

[88] Finlayson, A. M.—The pyritic deposits of Huelva, Spain. Econ. Geol., 5, 357-372, 403-437 (1910).

[89] Mineral Deposits, 602 et seq. (1913).   [90] Loc. cit., 114.

mens include representative suites of most of the important deposits except those from Norway. It is true that pyrite occurs in some of the Sudbury mines; we have examined several of these occurrences, and have found the pyrite to be distinctly later than the nickel-copper sulfids.

Later mineralization seems to be more prominent at the Worthington mine than at any other locality in the Sudbury district. Walker [91] says: "It is difficult to avoid the conclusion that the ores as they are now found at the Worthington mine have been subject to rearrangement by aqueous agencies since the solidification of the rock and sulfids from the original magma." Pyrite is evidently not a typical mineral at Sudbury. Browne,[92] for example, says: "Practically speaking, there is no pyrite, marcasite, or any other sulphide (he has previously mentioned pyrrhotite, chalcopyrite, and pentlandite) found in the great ore bodies."

Pyrite is entirely absent in the ores from the Engels mine. This might be attributed to lack of sufficient iron to combine with sulfur; but in the Namaqualand ores there was sufficient iron to form pyrrhotite, yet pyrite is entirely absent.

As a matter of fact, pyrrhotite seems to take the place of pyrite in the magmatic ores. The work carried on [93] at the Geophysical Laboratory, proving that pyrrhotite is the iron sulfid stable at high temperature, may explain this. Pyrite has usually been considered a persistent mineral.[94] Its maximum development, however, is reached in hydrothermal deposits, and it is less important in contact deposits and in those formed at low temperatures by meteoric water. True, pyrite is common in igneous rocks; but in the great majority of cases it has been introduced by hydrothermal solutions. There are a few authentic cases[95] of pyrite in fresh unaltered igneous rocks, and so its occurrence as a magmatic mineral can not be absolutely denied; but it seems certain that it is of minor importance as a magmatic mineral.

[91] Walker, T. L.—Certain mineral occurrences in the Worthington mine, Sudbury, Ontario, and their significance. Econ. Geol., 10, 542 (1915).

[92] Browne, D. H.—Notes on the origin of the Sudbury ores. Econ. Geol., 7, 468 (1906).

[93] Allen, E. T., Crenshaw, J. L., and Johnston, J.—The mineral sulphides of iron. Am. Jour. Sci. (4), 33, 169-236 (1912).

[94] Lindgren, W.—The relation of ore-deposition to physical conditions. Econ. Geol., 2, (1907).

[95] Lindgren, W.—The gold-quartz veins of Nevada City and Grass Valley districts, California. 17th Ann. Rept. U. S. Geol. Surv., pt. II, 66 (1896).

Spurr, J. E., and Garrey, G. H.—Prof. Paper no. 63, U. S. Geol. Surv., 388 (1908).

MAGNETITE-ILMENITE DEPOSITS

Geologists apparently find little difficulty in accepting the titaniferous magnetites as magmatic deposits, probably because magnetite is a common accessory mineral of igneous rocks, but appear to be more cautious in regard to the sulfid deposits. We have examined a number of examples of the magmatic iron ores in "basic" rocks, and find that the relations of the ores to silicates are similar in all respects to those of the sulfid deposits. In the latter the sulfids may fail locally and the ores become identical with the low-grade titaniferous iron ores. The sulfid rich and sulfid free magnetite deposits alike develop subhedral crystals within the silicates; and the larger anhedral crystals cut, surround, and replace the silicates. They are often unaccompanied by secondary alteration products, and where these are present, they are later than the ores.

In order to check our incomplete study, we have examined the literature of these deposits, and find, contrary to the general impression, that the majority of those who have made a careful microscopic study of the ores conclude that they are later than the silicates. As a result of a similar review of the literature, Lindgren [96] states: "In these differentiated magmas ilmenite and magnetite have, as a rule, crystallized after the silicates." Berg [97] notes: "Die grosseren Erzmassen sind stets Anhaufungen des jungeren Erzes, umschliessen also einzelne Krystalle von Hypersthen, Diallag, Olivin, und Feldspat."

OTHER MAGMATIC IRON ORES

It is not our purpose to take up the remaining types of iron ores for which a magmatic origin has been suggested, the origin of some of which is obscure on account of the metamorphism they have undergone. In general, those deposits related to the less "basic" rocks, such as the Kiirunavaara type,[98] and certain of the Adirondack ores,[99] show evidence of increasing activity of mineralizers, with the occasional development of pneumatolytic and allied minerals, and a tendency to migrate further out from the mother rock.

In general, iron oxids, like the magmatic sulfids, appear to be concentrated, not during the early magmatic stages, but during the later stages and under the influence of mineralizers.

---

[96] Mineral Deposits, 749.　　　　　　　[97] Loc. cit., 102.

[98] Igneous rocks and iron ores of Kiirunavaara, Luossavaara, and Tuollavaara. Econ. Geol., 5, 696-718 (1910).

[99] Newland, D. H.—On the association and origin of the non-titaniferous magnetites in the Adirondack region, 2, 763-773 (1907).

CHROMITE DEPOSITS

Of the remaining ore-minerals concentrated as magmatic segregations, nickel-iron, gold, and platinum, do not occur in sufficient amounts to be considered as ores.

Chromite in unaltered rock from Norway, has been described by Vogt,[100] and is considered by him to be the oldest mineral of the rock, on account of its occurrence as sharp octahedra. Examination of his figures shows irregular as well as euhedral forms; and the alternate hypothesis that the chromite octahedra are formed at a late stage, is worthy of consideration.

Vogt states: "The chromite deposits in peridotite show accordingly the same geological, petrographical and morphological characteristics as those of titaniferous iron in gabbro, and the general genetic statements afterwards enumerated in connection with the titaniferous iron deposits hold good also for the occurrence of chromite." [101]

[100] Beiträge zur genetischen Classification der durch magmatische Differentiationsprocesse und durch Pneumatolyse entstanden Erzvorkommen. Zeit. f. prakt. Geol., Jahrg. 1894, pp. 384-393.

[101] Ore Deposits: Beyschlag, Vogt, and Krusch; Translated by Truscott, (1914). 246.

PART III.
SUMMARY AND CONCLUSIONS

CRITERIA FOR THE RECOGNITION OF MAGMATIC ORES

Any discussion of the criteria by which the magmatic ores may be separated from other types of ores must be somewhat tentative in nature, as long as the latter have not received detailed comparative study.

The fact that the ores migrate only a short distance, if at all, into the adjoining rock, distinguishes them from contact deposits formed chiefly in the intruded rock, and from many other high-temperature deposits.

Comparison, then, must be made with ores that occur in, and as an integral part of, igneous rocks.

The fact that the only high-temperature alteration mineral present in appreciable amount is hornblende, which we believe is formed by magmatic processes, and that tourmaline and garnet are only occasionally developed, distinguishes these deposits from those in which destructive pneumatolytic and contact action is prominent.

The fact that hydrothermal alteration of the silicates is often minor in amount, is invariably later than the ores, and generally is not accompanied by the deposition or migration of ores, differentiates these ores from those of hydrothermal origin.

The absence of silicification and albitization, and the only occasional development of considerable amounts of sericite, separate these deposits from moderate- and high-temperature ores in igneous rocks, such as the copper ores of Bingham and Butte and the gold ores at Treadwell, Alaska, etc.

There is definite order of succession of the magmatic ore-minerals, as follows: magnetite, hematite, pyrrhotite, pentlandite, chalcopyrite, bornite. Any change in this order is due to rearrangement subsequent to the magmatic period.

Pyrite is not a typical magmatic mineral; if present in a magmatic ore, it is introduced later than the magmatic period.

SUMMARY OF THE CHARACTERISTICS OF MAGMATIC ORES

Contrary to general opinion, the magmatic sulfids are formed by replacement of the silicates after the solidification of the igneous rock. Notwithstanding this, we retain the term "magmatic ore deposits" for the types of ores described in this paper, because they have a close genetic relation to the intrusive rock in which they occur, and because they are formed within the magmatic stage as defined by us. Regardless of any theory as to their genesis, however, these deposits have definite and easily recognizable characteristics, which distinguish them from all other types of ore deposits.

The characteristics of the magmatic sulfid ores as brought out by our study may be summarized as follows:

(1) They occur in subsilicic rocks of the norite, gabbro, peridotite, or related families.

(2) In most occurrences the containing rock is either dominantly subsilicic, with minor amounts of complementary persilicic differentiates (Insizwa), or occurs as lenses of mafic rock in a large granitic intrusion (Golden Curry).

(3) The subsilicic rock occurs generally as small dikes, sills, or stocks, and rarely as a large laccolith (Sudbury).

(4) In most cases the ore-bearing rock has undergone marked differentiation, and the differentiated portions are sharply separated and do not grade into each other (Ookiep).

(5) The ore may occur in any variety of rock produced by differentiation, but in any one locality the ore shows marked preference for certain types of rock and occurs sparingly in others (Ookiep).

(6) Pegmatite and aplite dikes often cut, and are therefore later than, the ore bodies (Erteli).

(7) The ore is generally segregated at the margins of the intrusives, but occasionally occurs as lenses or tabular ore shoots well within the intrusive magma (Engels). In sills, the ore is usually at the base of the intrusives (Insizwa); in dikes, it often is formed along the footwall (Sohland), or as columnar or irregular shoots occupying the entire width of the dike (Copper Cliff).

(8) Very often the ore is concentrated in those portions of the intrusive which have suffered brecciation during intrusion (Sudbury).

(9) The ore migrates only a short distance into the adjacent rock, from a few inches to a few score of feet at most (Sudbury).

(10) The "offset" deposits, formed in dikes at some distance from the main intrusive, are accompanied and cut by veins carrying ore and gangue minerals of hydrothermal origin (Sudbury).

(11) There is no essential distinction between the sulfid group and the magnetite-ilmenite group as to the origin or the relation of the ores to the silicate minerals. In all the magmatic ores examined, ore deposition takes place at the close of the magmatic period.

(12) There are two general classes of the magmatic sulfid ores: (*a*) pyrrhotite-chalcopyrite deposits, and (*b*) chalcopyrite-bornite deposits.

(13) The so-called pyritic intrusive ores are not magmatic.

(14) The principal ore-minerals of the magmatic period include magnetite, ilmenite, hematite, pyrrhotite, pentlandite, chalcopyrite, and bornite.

(15) Pyrrhotite and bornite have not been found together in magmatic ores.

(16) Pyrite is not a typical magmatic mineral.

(17) The ore-minerals are formed at a late magmatic stage by a partial replacement of the silicate minerals. The ores surround, embay, and cut all the earlier silicates. They penetrate the cleavable minerals. They occasionally occur as sharp veinlets which lead out from the larger sulfid masses. Selective replacement is shown by the preservation in the ores of an original graphic texture of the rock.

(18) There is a definite order of formation of the principal magmatic ore-minerals. This order is as follows: magnetite-ilmenite (intergrowth), hematite, pyrrhotite, pentlandite, chalcopyrite, and bornite.

(19) There is evidence of the replacement of one magmatic ore-mineral by another.

(20) Euhedral magnetite and probably other minor accessories occurring in euhedral crystals, such as apatite, zircon, titanite, etc., are also formed at a late magmatic stage.

(21) There is clear evidence of the magmatic alteration of pyroxene to hornblende prior to the introduction of the ore-minerals.

(22) Hydrothermal alteration, altho seldom lacking in magmatic ores, is relatively insignificant, and is distinctly later than the magmatic ore period. The silicates of the hydrothermal period include tremolite, anthophyllite, sericite, chlorite, and serpentine. These secondary silicates often replace the magmatic ore-minerals in veinlets and in sharp

lath-shaped crystals without causing any change or migration of the ore-minerals.

(23) The attention given these alteration products in order to determine their relative age, may have given an erroneous impression as to their relative abundance. In most of the ores studied they are present only in minor amounts. Many of the magmatic ores are as free from alteration products as the average unmineralized igneous rock.

(24) However, a minor amount of rearrangement, consisting in the production of microscopic crystals of pentlandite and chalcopyrite of the second generation, has been detected in the pyrrhotitic ores. In the chalcopyrite-bornite ores there has been some migration, resulting in the formation of minor amounts of covellite, chalcocite, and chalcopyrite of the second generation. This alteration is only prominent where there has been an abnormal development of sericitization.

(25) The rôle of mineralizers in magmatic differentiation has not been sufficiently emphasized. The crystallization of the early formed minerals in the magma involves the complementary process of the "squeezing out" of the residual fluid. This process is not merely a mechanical one, but is also due to gaseous extraction.

(26) The typical magmatic deposits, unaccompanied by high-temperature alteration, with the exception of magmatic hornblende, are chiefly developed in "basic" rocks. Ore deposits genetically related to persilicic rocks show intense rock alteration, probably the result of mineralizers more "active" than those accompanying the subsilicic rocks.

(27) There is a parallelism between the various groups of high-temperature deposits, of which the magmatic ores are one division. In all groups, high-temperature silicates precede the introduction of the ore, and hydrothermal stages follow. In contrast with the magmatic deposits, the non-magmatic ores are characterized by a complex set of pre-mineral silicates, and the hydrothermal stage is generally the most important period of ore introduction.

(28) The temperature at which the introduction of ore-minerals is initiated is about the same for all the high-temperature deposits; probably not higher than 300-400° C.

(29) Gradations between the typical magmatic ores and other high-temperature deposits are shown by the local development of garnet and tourmaline. Further, the gradations of these into intermediate-temperature and low-temperature deposits is strong evidence of the magmatic origin of ore deposits in general.

(30) The following orderly series of events is recognized in magmatic deposits:

(*a*) Crystallization of primary silicates;

(*b*) The development of hornblende and biotite, and occasionally tourmaline and garnet, as magmatic alteration products;

(*c*) The introduction of the ore-minerals;

(*d*) A small amount of rearrangement of the ores and the development of secondary silicates by hydrothermal solutions.

Department of Geology,
Leland Stanford Junior University,
     October, 1916

EXPLANATION OF PLATES AND METHODS OF PREPARING PHOTOGRAPHS AND SECTIONS

The photographs are of two types: thin sections and polished sections. The thin sections are necessary for the identification of the silicates; the polished sections for the identification of the ore-minerals,

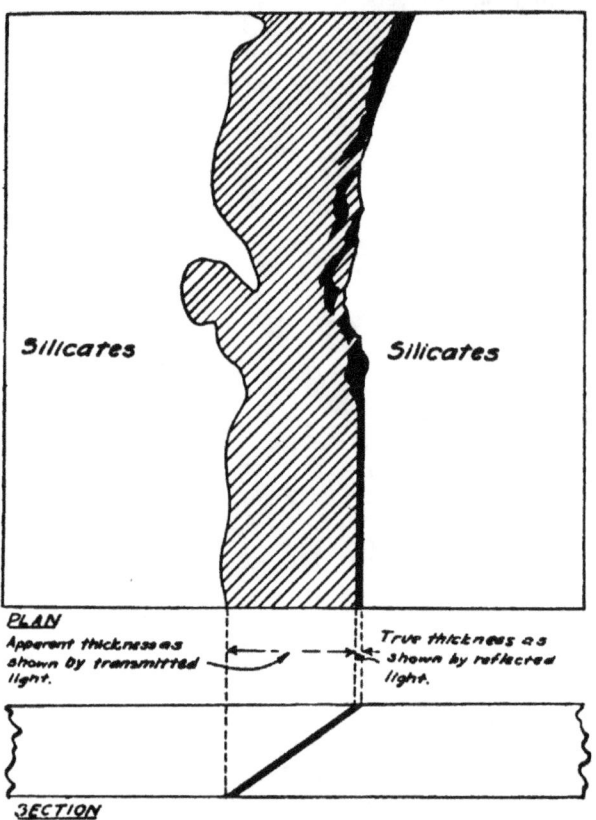

Fig. 6A.   Drawn from a photograph of a polished thin section of ore from the Stobie mine, Sudbury. White areas are silicates. Lined area is opaque band as viewed by transmitted light. Black band is veinlet as shown by reflected light. (x 175)

and to show the relations of the various minerals to each other. In examining the photographs note that the ore-minerals are black and the silicates light in thin sections, while the silicates are dark gray and the ore-minerals white or light gray in the polished sections.

In taking the photographs we have used Wratten panchromatic plates, selecting in each case the color screen best adapted to bring out all the minerals in the polished section.

In the microscopic investigation we examined (1) thin sections with the polarizing microscope; (2) polished sections with the reflecting microscope; and (3) polished thin sections (combination sections) with both transmitted and reflected light.

It is now generally admitted that the examination and determination of the opaque minerals in thin sections is at best unsatisfactory, and that only the grosser relations of the opaque minerals to each other can be made out, and that those details requiring high magnifications are lost entirely.

In addition to the above, we have found that the relations of the opaque minerals, as a group, to the transparent minerals, are shown more accurately in polished surfaces than in thin sections, and that the relations of the transparent silicates to each other are sharper when viewed with the reflecting microscope than with the polarizing microscope, altho the identification of the transparent minerals must be made with the latter.

The reason for this is simple. In examining a thin section, one looks thru a certain thickness of rock, and not at a single definite surface. All contacts between the opaque and transparent minerals that are not perpendicular to the surface of the slide are widened so that the black area represents the projection on the surface of the section of all the opaque material contained in the slide. A thin section does not show, therefore, the relation of the opaque and the transparent minerals at one definite level as do the polished sections. Fig. 6a, page 74, was drawn from a photograph of a polished thin section viewed under both transmitted and reflected light. The ruled band represents the shadow cast by the opaque material as viewed with transmitted light, and the thin black line at the margin of the band shows the true thickness of the veinlet as registered by the light reflected from the polished surface.

In order to take advantage of the reflecting microscope for this work, it was necessary to develop special methods of polishing in order to give a satisfactory brilliant surface to the silicates as well as to the hard and soft opaque minerals. As a matter of fact, the polishing of all the different constituents in a single section of complex ore is an art, and, up to date, only crude methods of polishing have been described in the literature.

Our methods, in brief, consist in grinding on a glass plate a plane surface with the minimum of pits, using successively finer grades of carborundum, and then "sixty-minute" carborundum, and finally a still finer carborundum prepared in our laboratory.

The final polishing is done with fine cambric on rapidly rotating wheels (1700-3400 r.p.m.). Silicates, magnetite-ilmenite, pyrite, etc., are polished by carborundum, prepared by grinding the "sixty-minute" grade two weeks in a ball mill, and then floating off the finest product for this purpose. Chrome oxid is used to polish sulfids of average hardness, and aluminum oxid for the softer sulfids. These powders are applied one after the other, or as mixtures, depending upon the character of the specimen. To develop a polish of the highest brilliancy on the sulfids— for example to bring out pentlandite enclosed in pyrrhotite—aluminum oxid is used on the finest billiard felt.

In most cases we find it satisfactory to use one side of a sawed specimen for the polished section and the adjacent surface for the thin section. In some cases a thin section is given a high polish, which acts like a cover-glass in giving clear images with transmitted light, and also gives brilliant reflections with reflected light.

PLATE I

PHOTOGRAPHS OF POLISHED SPECIMENS OF SUDBURY ORES

FIG. 7

Rich ore (pyrrhotite and pentland-ite) from Copper Cliff mine, show-ing unreplaced silicates [dark areas] of the granitic rock.

Nat. size

FIG. 8

Ore from the Creighton mine showing chalcopyrite veinlets.

x 1 2/3 diameters

FIG. 9

Nodule of "basic" material in granitic rock which has been largely replaced by pyrrhotite [light gray]. Creighton mine.

x 87/100

FIG. 10

Pyrrhotite [light gray] impregnat-ing and replacing "greenstone schist." Garson mine.

x 8/10

Fig. 7

Fig. 8

Fig. 9

Fig. 10

PLATE II

Photomicrographs of thin sections of lean ore from the Stobie mine, Sudbury

Fig. 11

Typical norite with euhedral magnetite [black], hypersthene [gray], plagioclase [white].

Magnification x 36 diameters.

Fig. 12

Ore-minerals; magnetite, pyrrhotite [black], surrounding and replacing hypersthene [gray]. A veinlet of chlorite [ch] cuts magnetite. In the upper portion of the photograph the hypersthene is fresh, and in the lower part it is altered; and yet the ore shows the same relation to the silicates in the fresh and altered portions.

x 30

Fig. 13

A veinlet of chlorite [ch], cutting the sulfids, chalcopyrite and pyrrhotite [black], and the silicates.

x 36

Fig. 14

Tremolite [tr] crystals cutting— not residual in—chalcopyrite [black], and projecting from altered hypersthene [gray].

x 119

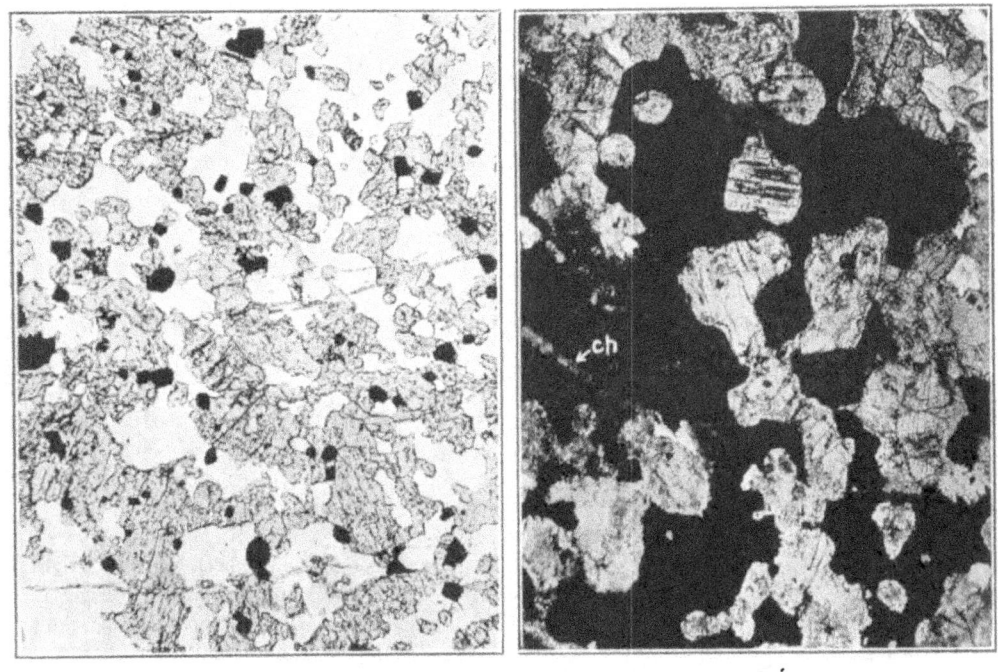

Fig. 11   Fig. 12

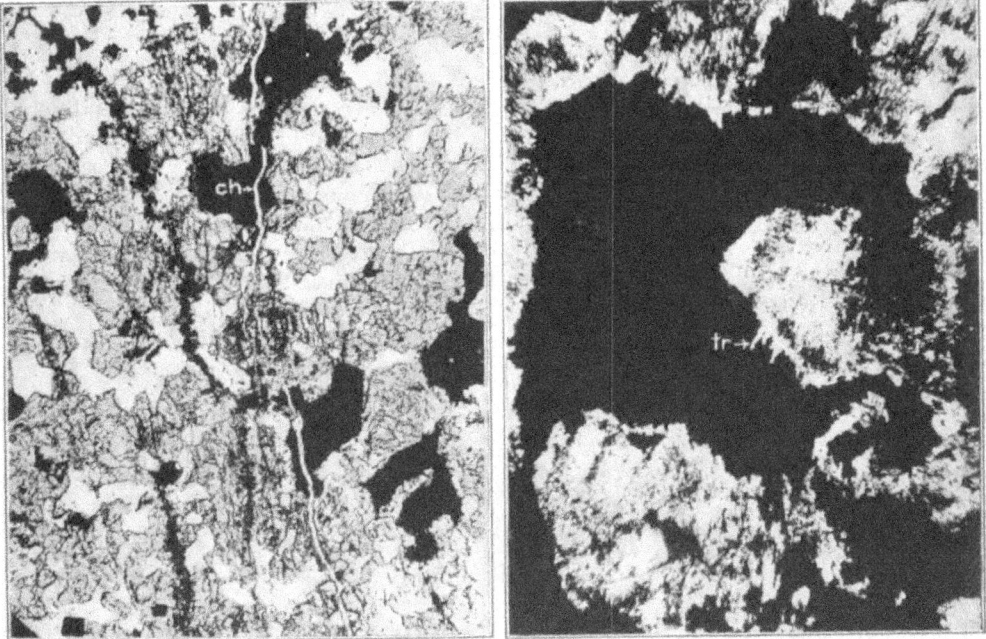

Fig. 13   Fig. 14

PLATE III

PHOTOMICROGRAPHS OF POLISHED SECTIONS OF ORES FROM THE STOBIE
MINE, SUDBURY

FIG. 15

Lean ore. Chlorite veinlets cutting ore-minerals; magnetite, pyrrhotite, chalcopyrite [light], and the silicates.

x 11

FIG. 16

Plagioclase [dark], uralite [light gray], and sulfids [white]. Figs. 17 and 18 are higher magnifications from the same field.

x 10

FIG. 17

Veinlet and crystals of tremolite [gray] cutting pyrrhotite.

x 270

FIG. 18

Pyrrhotite [p] and pentlandite [pn]$_1$ cut by irregular veinlets of secondary silicates [ss]. A second generation of pentlandite (?) [pn]$_2$ grows out in tufts from these veinlets. The secondary migration of ore-minerals is therefore insignificant in amount.

Aggregates of tremolite crystals on the right of the photograph.

x 165

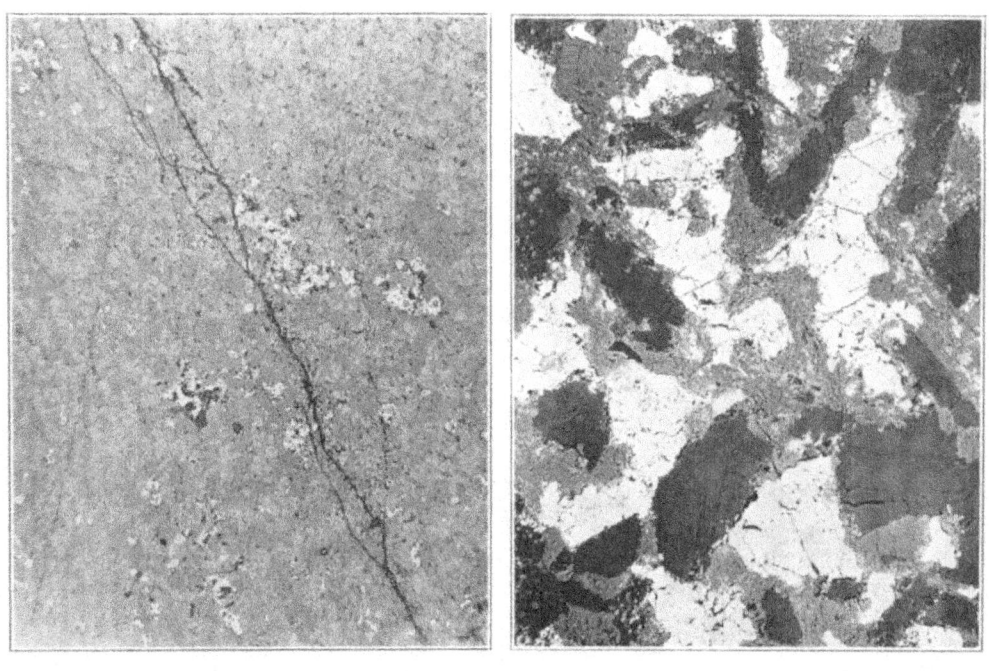

Fig. 15   Fig. 16

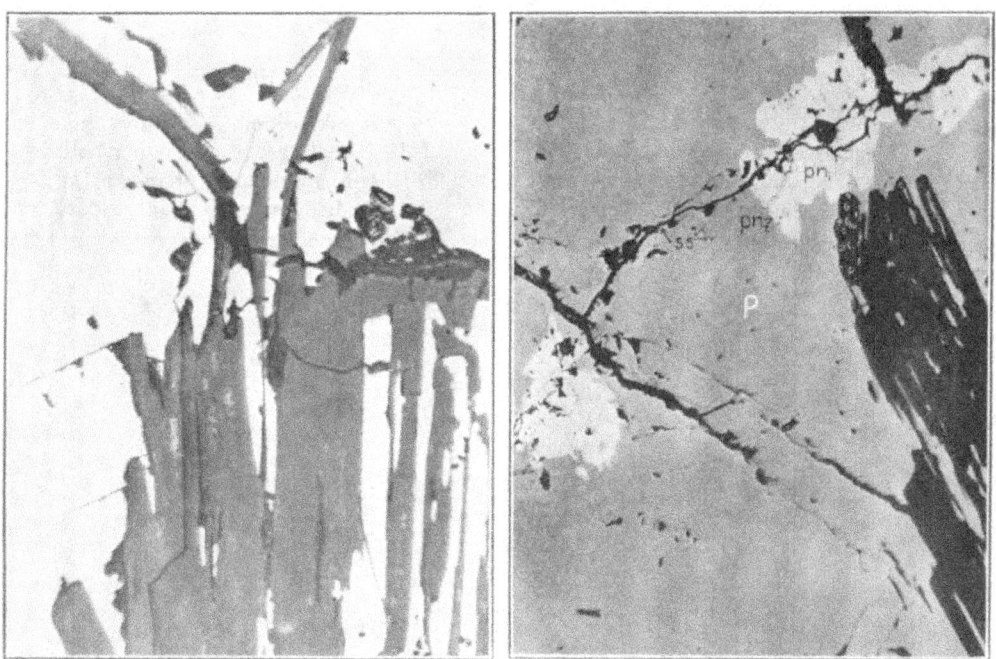

Fig. 17   Fig. 18

PLATE IV

Fig. 19

Polished section of rich ore from the Stobie mine. Irregular areas and veinlets of chalcopyrite [white] replacing garnet [g] and other silicates.

x 16

Fig. 20

Copper Cliff mine. Thin section of ore in coarse granitic material. Ore-minerals [black areas], magnetite and pyrrhotite, cutting and replacing biotite.

x 38

Fig. 21

Polished section, Creighton mine. Crystallographic intergrowth of magnetite and ilmenite [parallel dark lines]. Specimen is not etched, but contrast is brought out by high polish.

x 790

Fig. 22

Polished section, Creighton mine. Typical rich ore showing pyrrhotite [p] with veins of pentlandite [pn] and residual magnetite [m] and silicates [s].

x 14

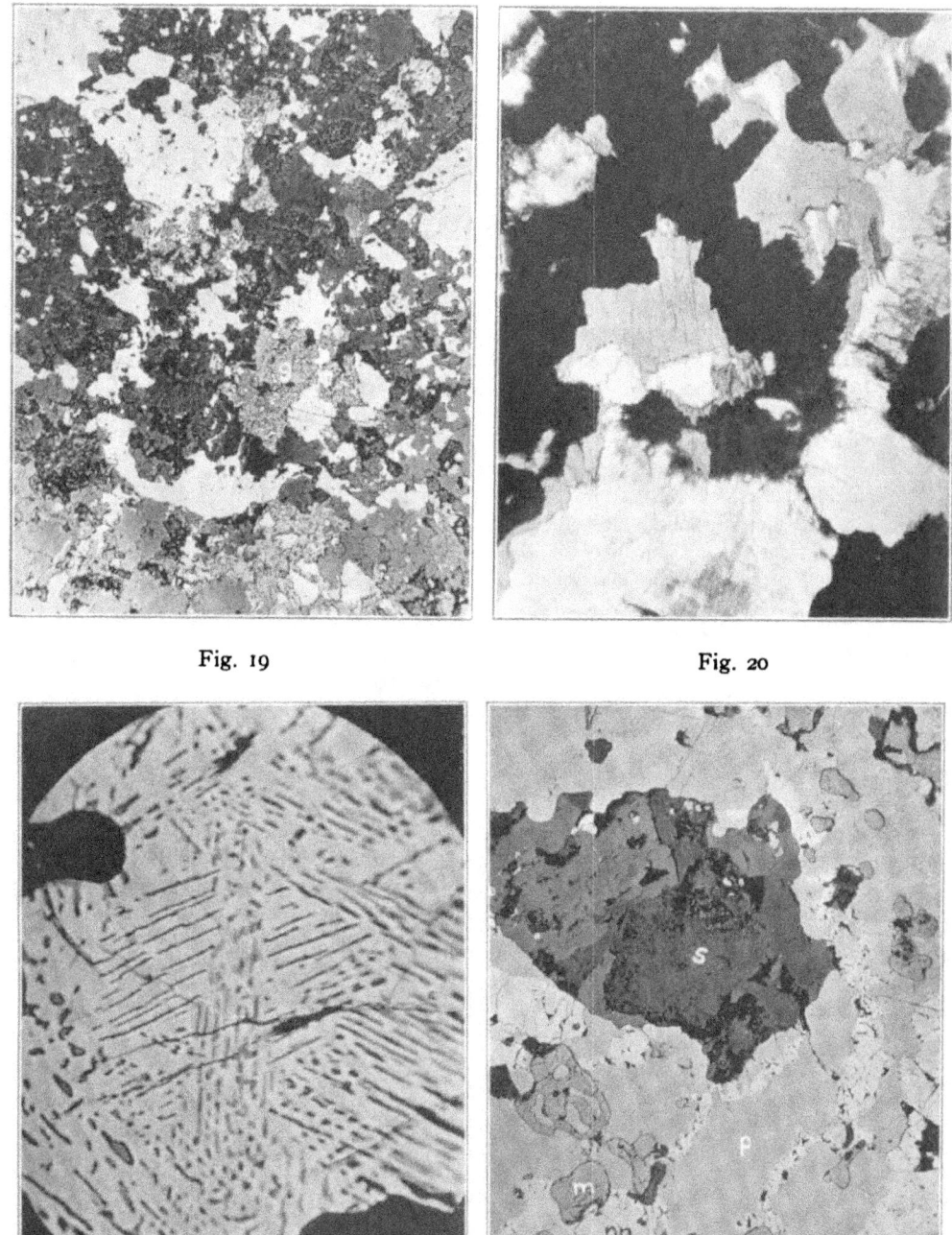

Fig. 19   Fig. 20

Fig. 21   Fig. 22

PLATE V

PHOTOMICROGRAPHS OF THIN SECTIONS OF ORE FROM SUDBURY

FIG. 23

Ore from the Creighton mine. Large, irregular ore masses [black] are chalcopyrite and pyrrhotite. Small subhedral black spots are chalcopyrite, probably a replacement of magnetite, as they contain residual specks of the latter. Ore-minerals cut through the silicates and replace them. Veinlets at the bottom of photograph are chalcopyrite.

x 28

FIG. 24

Creighton mine. Chalcopyrite veinlets [black] cut microcline, quartz, and biotite [dark gray], which are free from alteration products.

x 66

FIG. 25

Creighton mine. Sulfid veinlets, chalcopyrite and pyrrhotite, which extend out from a mass of sulfids [black], cut silicates [hornblende and plagioclase].

x 27

FIG. 26

Evans mine. Chalcopyrite veinlet cuts hornblende [ho], biotite [bi], and plagioclase [white]. Displacement, shown by hornblende crystal, has taken place along this veinlet. Beyond the field of the photograph chlorite cuts across the veinlet. [ap] apatite.

x 158

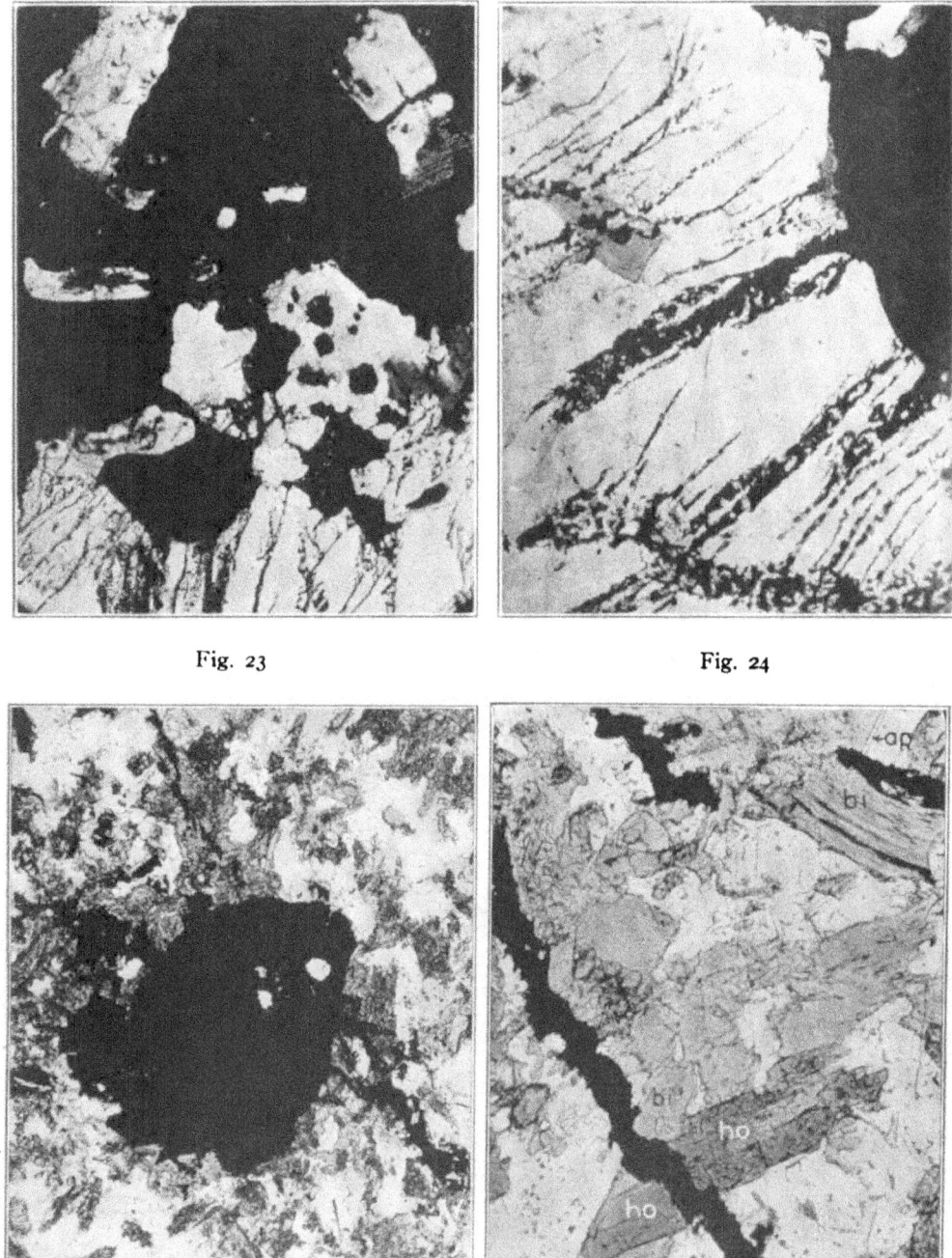

Fig. 23

Fig. 24

Fig. 25

Fig. 26

PLATE VI

PHOTOMICROGRAPHS OF POLISHED SECTIONS OF "ACID" ORE-BEARING ROCK
FROM THE CREIGHTON MINE, SUDBURY

FIG. 27

Ore masses, pyrrhotite and chalcopyrite, extend out into fine veinlets, chiefly chalcopyrite. Large masses replace the silicates. The veinlets cut the subgraphic intergrowth of quartz [q] and feldspar [f]

x 10

FIG. 28

Sulfids, pyrrhotite [p] and chalcopyrite [cp] cut magnetite [m] and silicates in anastomosing veinlets.

x 15

FIG. 29

Chalcopyrite [cp] extending from the main mass into veinlets as one generation of ore. No evidence of "rearrangement" or second generation of ore in veinlets. A mass of pentlandite [pn] enclosed in chalcopyrite.

x 99

FIG. 30

Sharp veinlets of chalcopyrite cut the graphic intergrowth of quartz [dark] and feldspar [light].

x 260

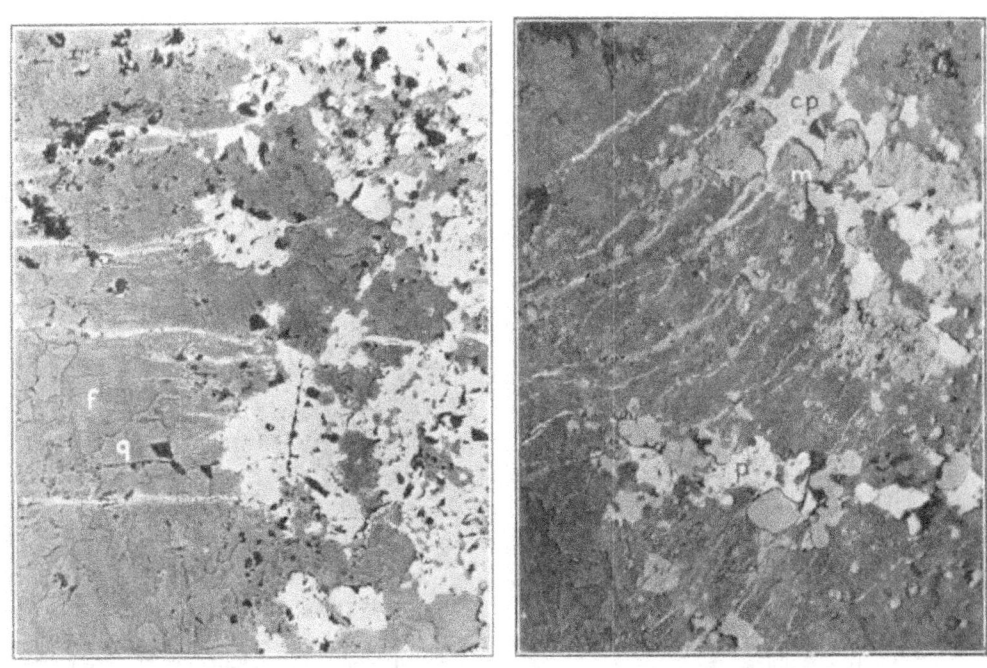

Fig. 27   Fig. 28

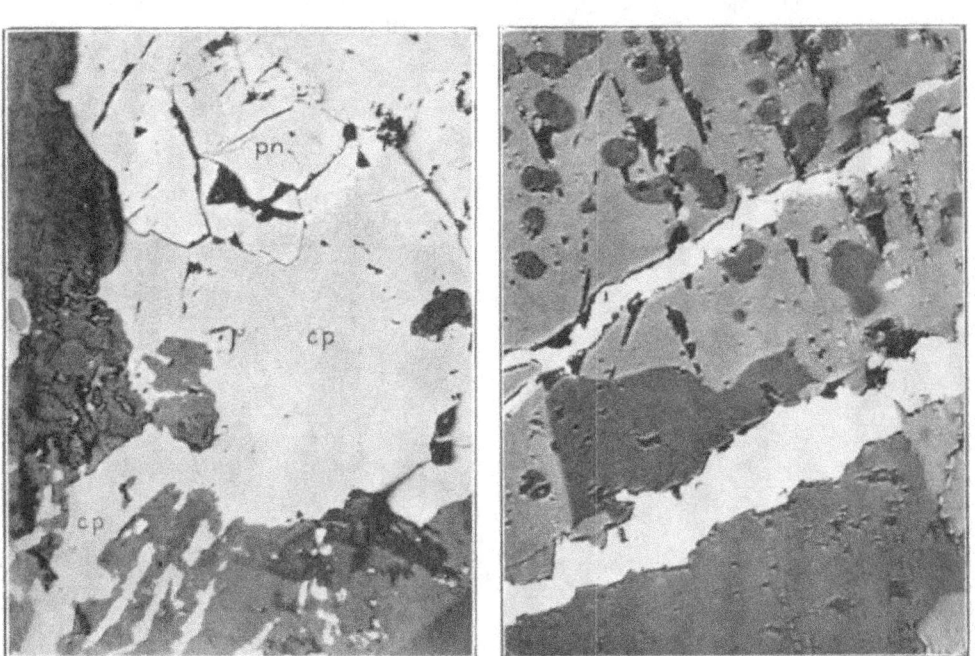

Fig. 29   Fig. 30

PLATE VII

PHOTOMICROGRAPHS OF POLISHED SECTIONS OF "ACID" ORE-BEARING ROCK FROM THE CREIGHTON MINE, SUDBURY

FIG. 31

Selective replacement by sulfids [white] of feldspar [light gray] in the feldspar-quartz intergrowth. Quartz is dark gray.

x 18

FIG. 32

Sulfids, pyrrhotite [p] and chalcopyrite [cp] and veinlike masses of pentlandite [pn]. In the center of the field chalcopyrite preserves the structure of the graphic intergrowth by selective replacement of the feldspar.

x 18

FIG. 33

Graphic intergrowth of quartz and feldspar. A small irregular replacement veinlet cuts the fine graphic intergrowth above. Large-scale graphic intergrowth below preserved by selective replacement of feldspar by sulfids.

x 71

FIG. 34

An irregular area of magnetite replacing silicates.

x 140

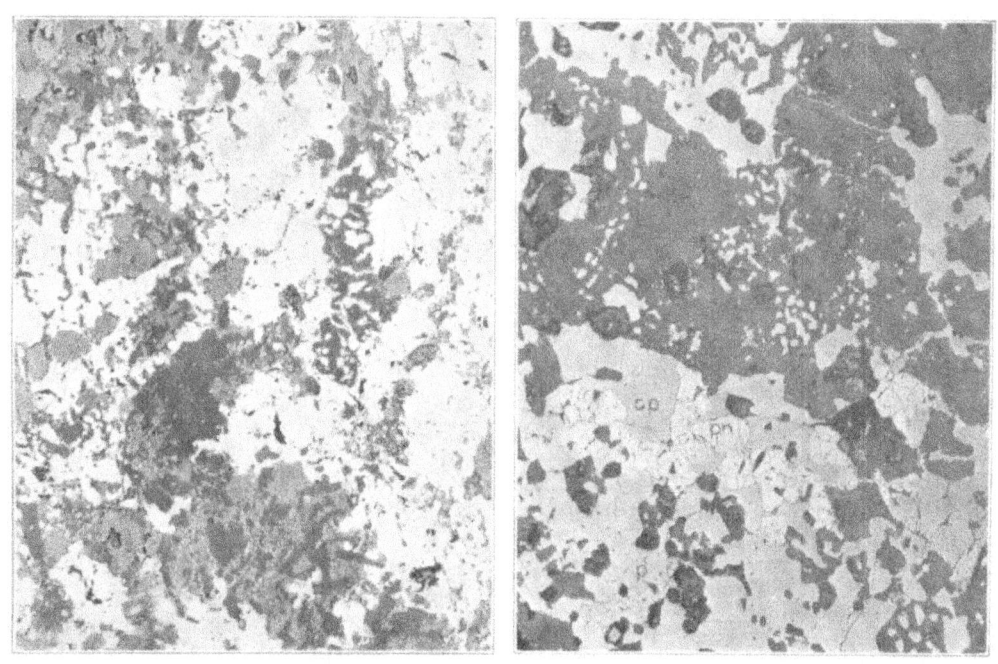

Fig. 31   Fig. 32

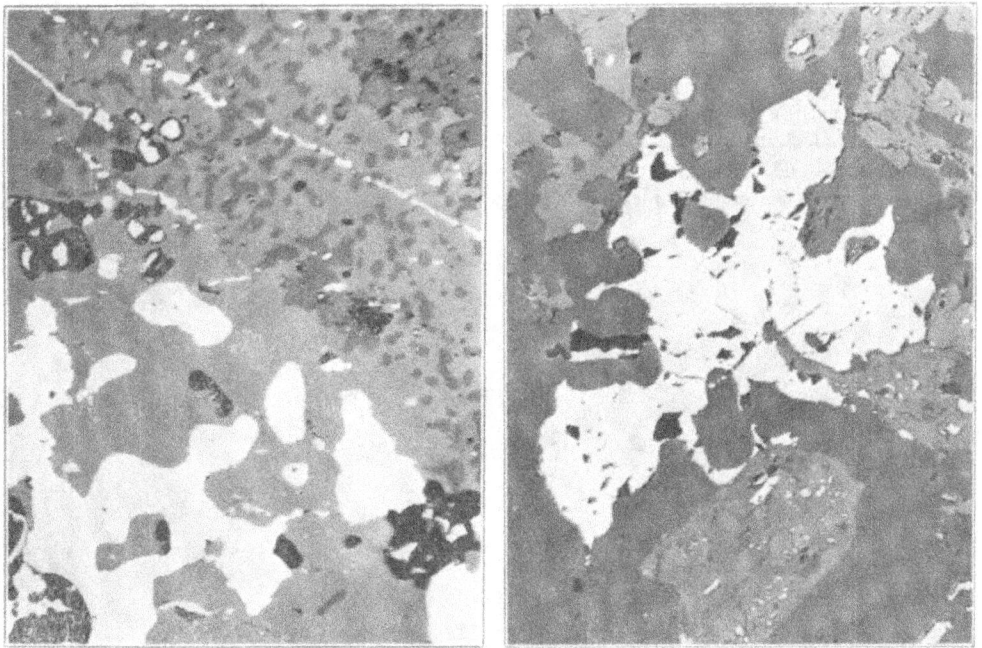

Fig. 33   Fig. 34

PLATE VIII

PHOTOMICROGRAPHS OF POLISHED SECTIONS OF ORES FROM SUDBURY

FIG. 35

Massive ore from the Creighton mine showing typical vein-like areas of pentlandite [pn], in pyrrhotite [p]. Contrast brought out by polishing, not by etching. Tiny brush-like crystals of a second generation of pentlandite [pn]₂ develop along veinlets and contacts. See Fig. 36.

x 18

FIG. 36

Massive ore from the Creighton mine. Second generation of pentlandite [pn] along a veinlet of chalcopyrite [cp], cutting pyrrhotite [p]. This rearrangement of ore-minerals is minor in amount, and can rarely be noted even under high powers. (Same specimen as shown in Fig. 35.)

x 670

FIG. 37

Massive ore from the Vermilion mine, showing pyrite veinlets [dark gray] cutting polydymite (?) [po] and chalcopyrite [cp]. Note that pyrite is definitely later than the typical magmatic ore-minerals in all the occurrences figured.

x 10

FIG. 38

Ore from the Worthington mine. Pyrite [white] with a reticulate structure, cutting gangue [black] and sphalerite [gray].

x 58

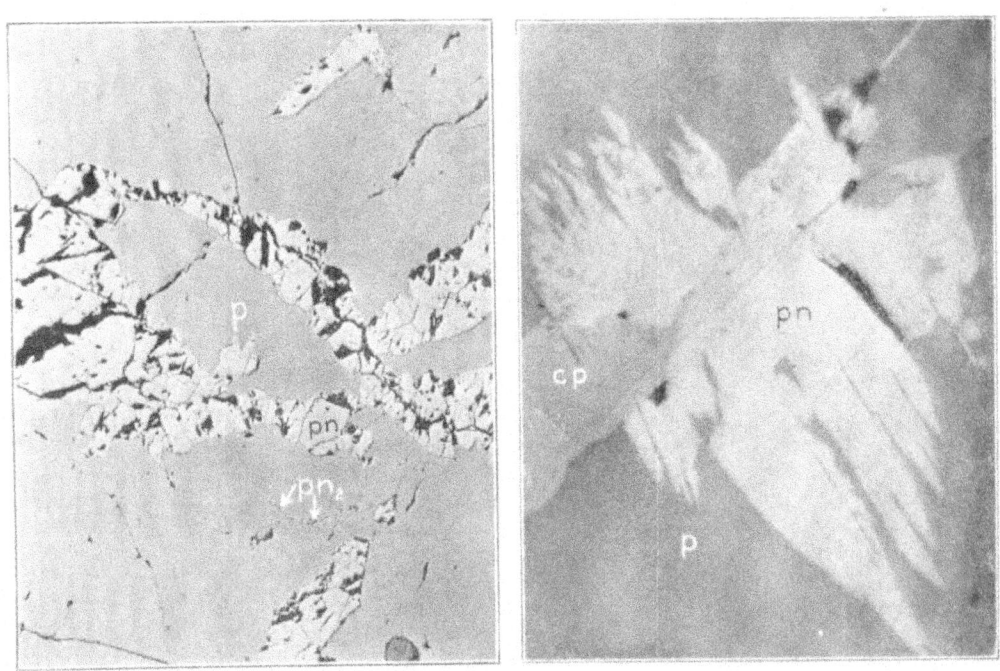

Fig. 35   Fig. 36

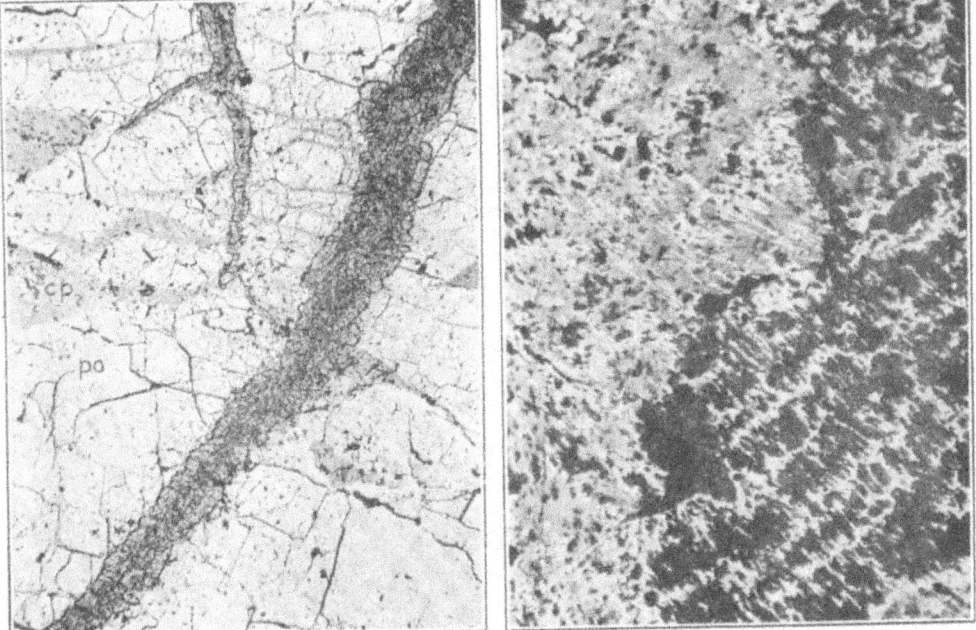

Fig. 37   Fig. 38

PLATE IX

PHOTOMICROGRAPHS OF POLISHED SECTION OF ORE FROM THE ALEXO MINE, NORTHERN ONTARIO, CANADA

FIG. 39

Shows the general relations of the ore-minerals [large white areas] to the serpentinized silicates. The original ore-minerals, chiefly pyrrhotite and pentlandite with subordinate chalcopyrite and magnetite, surround and embay the serpentine pseudomorphs after olivine.

A minor amount of the second generation of chalcopyrite has been concentrated by serpentinization within the silicates.

x 17

FIG. 40

Enlargement of a part of the field in the upper left-hand portion of Fig. 39. Shows serpentine pseudomorphous after olivine, surrounded by magmatic sulfids, the contacts of which are not modified by serpentinization. The latter, however, affects a secondary concentration of chalcopyrite developed as a border between two different types of serpentine.

x 70

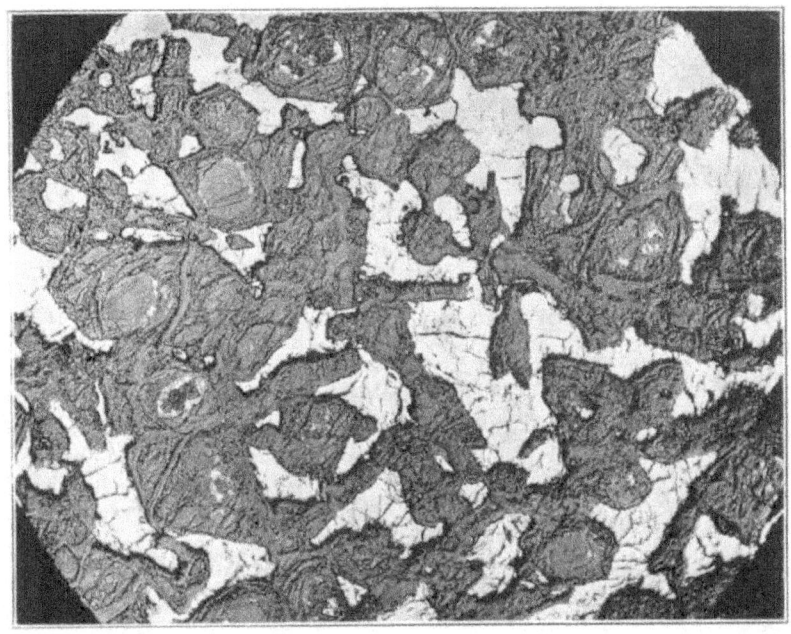

Fig. 39

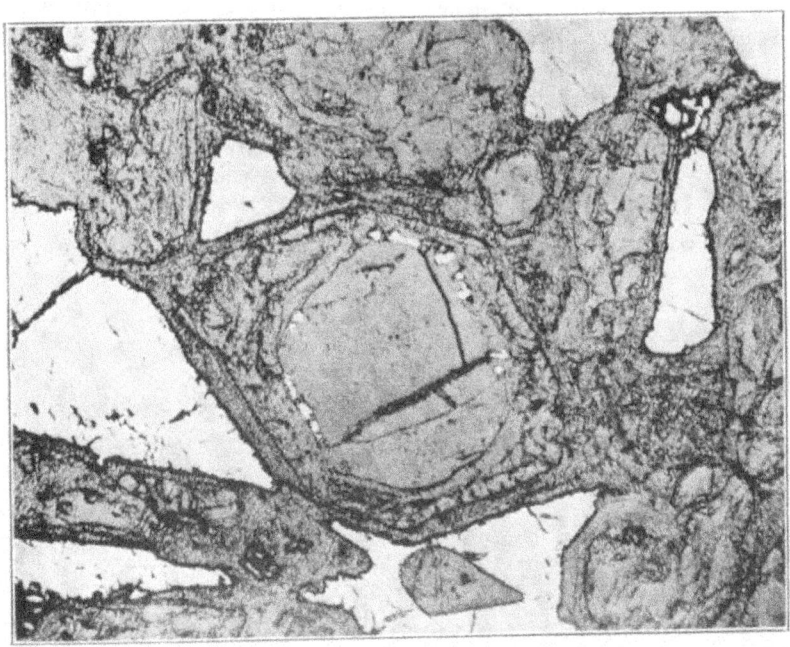

Fig. 40

PLATE X

PHOTOMICROGRAPHS OF ORE FROM THE ALEXO MINE, FROM THE SAME
POLISHED SECTION FIGURED IN PLATE IX

FIG. 41

Pyrrhotite [p], pentlandite [pn],
chalcopyrite [cp]₁, and magnetite
[m] cut by veinlets of serpentine.

x 44

FIG. 42

First [pn]₁ and second [pn]₂ generations of pentlandite in pyrrhotite [p]. The two generations can
be distinguished by a slight difference in relief and color. Serpentine
veinlets are later than the first generation pentlandite and closely connected with the second generation.
Pentlandite and serpentine tend to
develop along the crystallographic
lines of the pyrrhotite.

x 175

FIG. 43

Pyrrhotite [p] and pentlandite
[pn] of the first generation, cut by
a serpentine veinlet with a center of
magnetite [m], along which pentlandite of the second generation
[pn]₂ develops.

x 610

FIG. 44

Showing later generations of chalcopyrite [cp]₂ connected with different stages of serpentinization. A
large mass of pyrrhotite [p] is
shown near the top of the photograph. Small dots in center may be
a third generation of chalcopyrite.

x 183

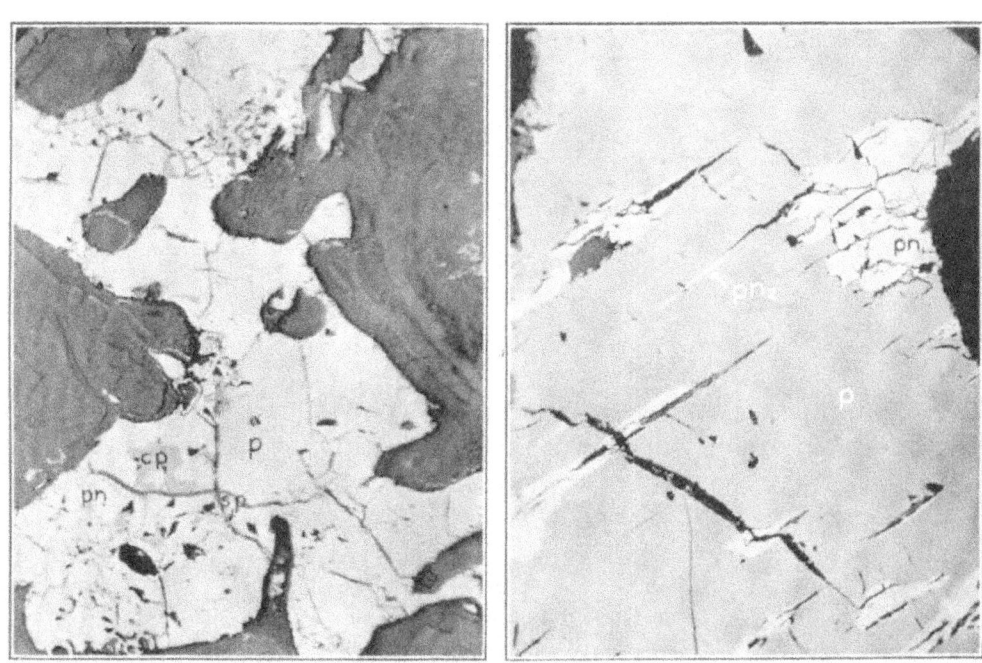

Fig. 41

Fig. 42

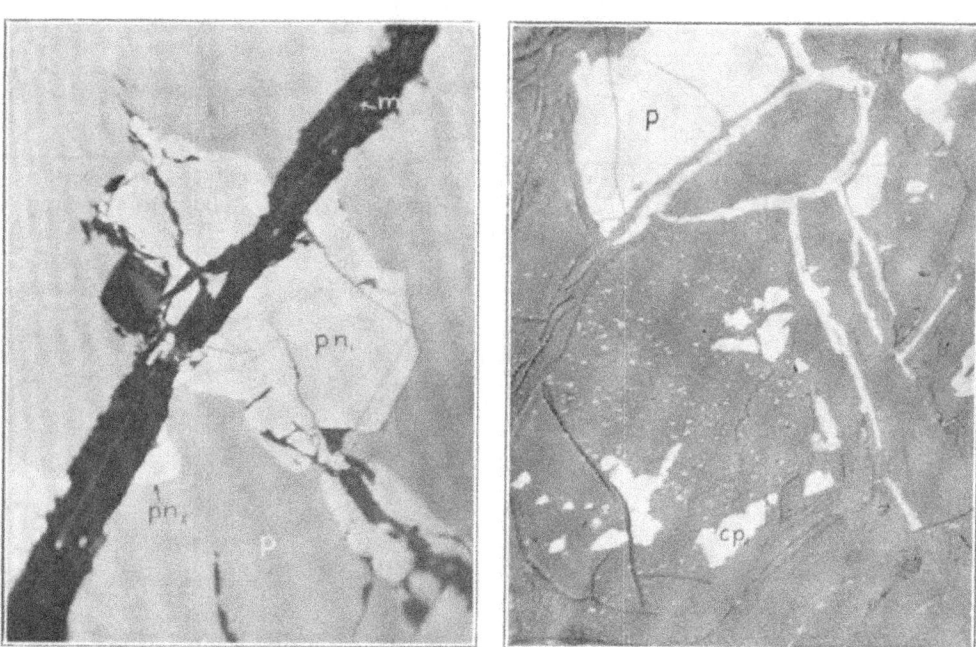

Fig. 43

Fig. 44

PLATE XI

PHOTOMICROGRAPHS OF THIN SECTION OF PYRRHOTITE-BEARING PYROXENITE FROM THE GOLDEN CURRY MINE, ELKHORN, MONTANA

FIG. 45

Sulfids [black], chiefly pyrrhotite with some chalcopyrite, surrounding and penetrating augite. White needles in the sulfids in the center of photograph are tremolite and darker patches to the right are hornblende.

x 22

FIG. 46

Sulfids replacing hornblende along cleavages.

[The white streak (lower center) is a crack in the slide.]

x 130

FIG. 47

Sulfids surrounding and penetrating pyroxene. A hornblende crystal [ho] in the center is bordered by tremolite [tr] needles which cut the ore-minerals.

x 90

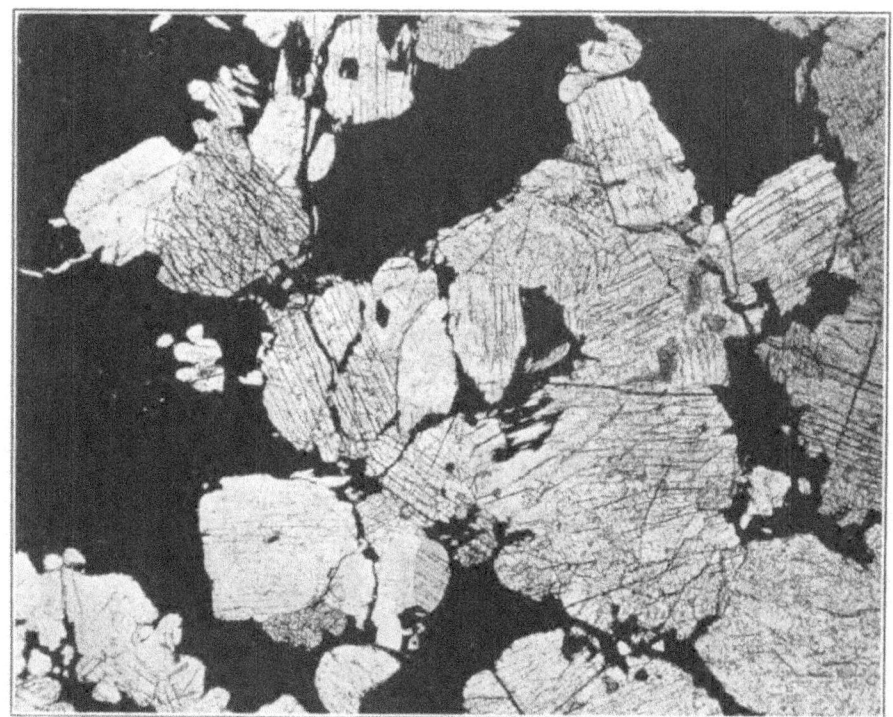

Fig. 45

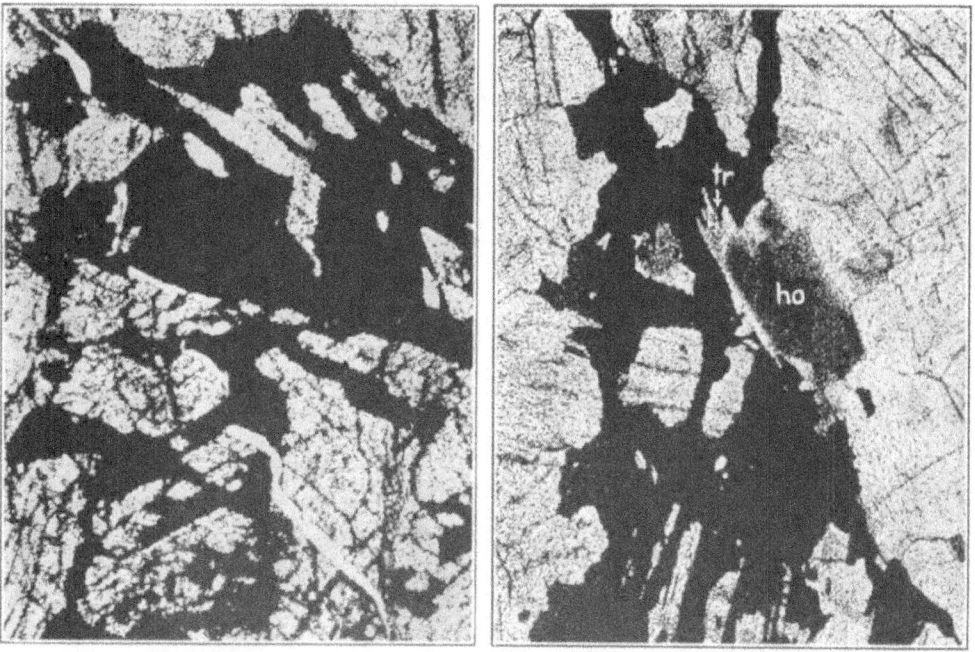

Fig. 46   Fig. 47

PLATE XII

Photomicrographs of polished sections from Evje, Norway, and Sohland, Germany

Fig. 48

Ore from the Flaad mine, Evje, Norway.

Pyrite [py] veinlet and detached crystals in pyrrhotite [p] and magnetite [m]. The veinlet is connected with alteration products. See fig. 49 for the relation of pyrite crystals to pyrite veinlets.

x 12

Fig. 49

Ore from the Flaad mine, Evje, Norway.

Pyrrhotite [p] cut by chalcopyrite [cp] which in turn is cut by pyrite [py] veinlets. The latter connect with subhedral and euhedral crystals. The euhedral crystals are shown in fig. 48.

x 300

Fig. 50

Ore from Sohland, Saxony.

Shows general relations. The ore-minerals are pyrrhotite [p], chalcopyrite [cp], and pentlandite [pn]. Uralite needles [u] penetrate into the sulfids. Other silicates [s] are hornblende and alteration products. A good example of extensive post-mineral alteration.

x 12

Fig. 51

Ore from Sohland, Saxony.

Uralite [u] needles cutting pyrrhotite [p] and pentlandite [pn]. The uralite (or tremolite) develops at the ends of the hornblende crystals.

x 116

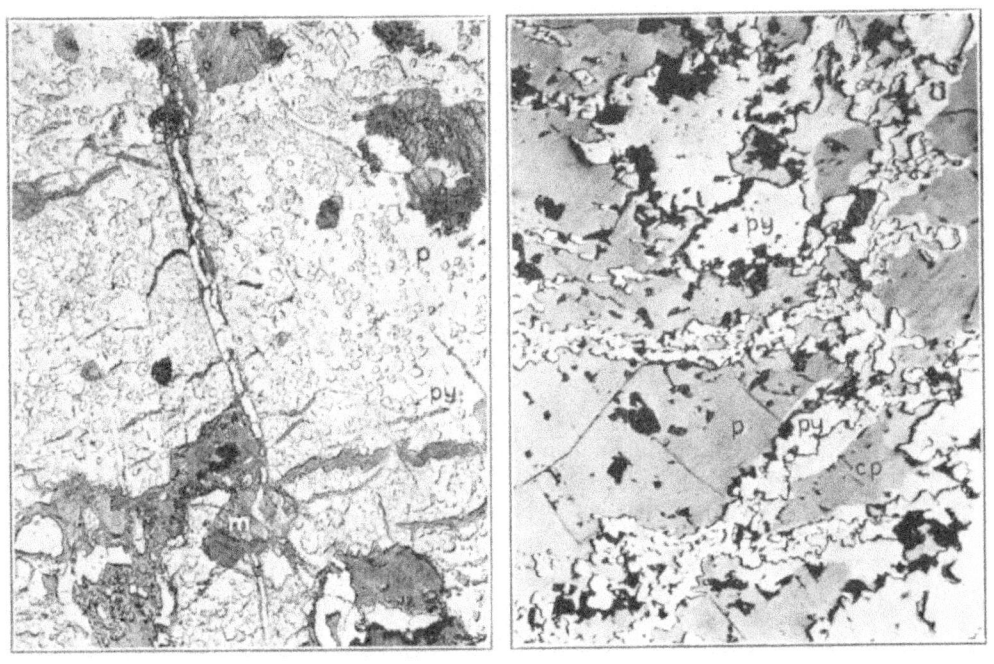

Fig. 48   Fig. 49

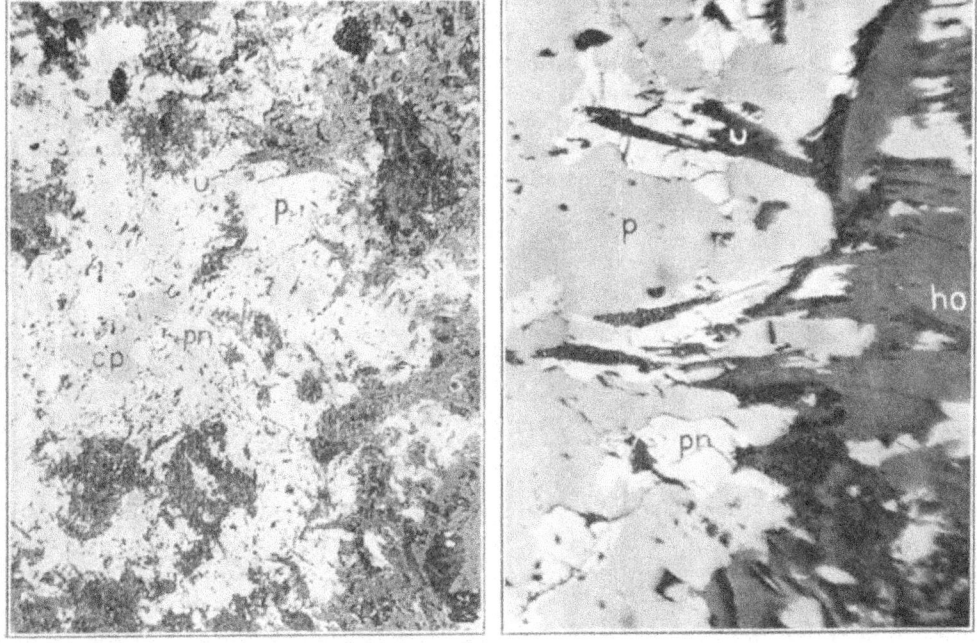

Fig. 50   Fig. 51

PLATE XIII

PHOTOMICROGRAPHS OF SECTIONS OF ORE-BEARING HYPERSTHENITE FROM NABABEEP MINE, OOKIEP, SOUTH AFRICA

FIG. 52

Thin section showing ore-minerals [black], magnetite and a little bornite, cutting and surrounding hypersthene.

x 74

FIG. 53

Thin section of hypersthenite showing euhedral crystals of magnetite within the hypersthene and anhedral ore-minerals, chiefly magnetite, connected with the "acid extract" [white]. The "acid extract" is plagioclase.

x 59

FIG. 54

Polished section of ore-bearing hypersthenite. The ore-minerals are magnetite [m] and bornite [b]. The general relations suggest that magnetite is partially replaced by bornite.

x 10

FIG. 55

Thin section showing anthophyllite [an] and talc (?) [ta] in pyrrhotite [black]. The sharp-pointed anthophyllite needles cut and are later than the ore-minerals.

x 435

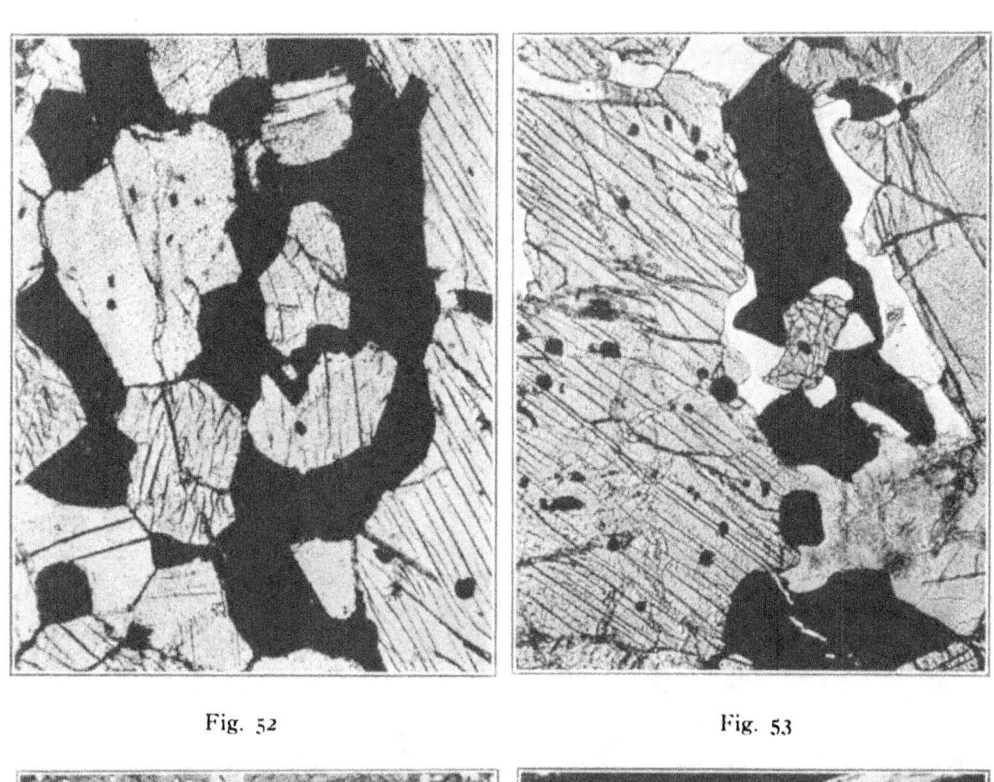

Fig. 52   Fig. 53

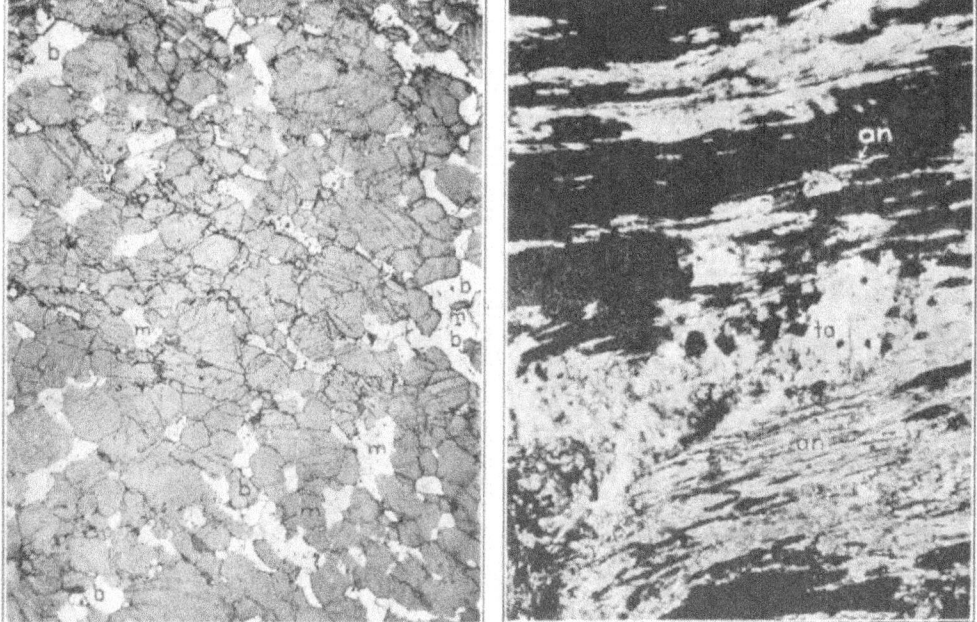

Fig. 54   Fig. 55

PLATE XIV

PHOTOMICROGRAPHS OF THIN SECTIONS OF ORE-BEARING NORITE FROM THE TWEEFONTEIN MINE, OOKIEP, SOUTH AFRICA

FIG. 56

Hypersthene [dark gray] and plagioclase [light gray] surrounded and cut by ore-minerals, magnetite and a little chalcopyrite. Euhedral magnetite [designated by arrow] is developed within the silicates. Silicates are free from any kind of alteration products.

x 17

FIG. 57

A higher magnification of the lower portion of fig. 56. Magnetite is anhedral along the boundaries of the silicates and euhedral [see arrow] within the plagioclase.

x 49

FIG. 58

Hypersthene [gray] and plagioclase [white] surrounded and replaced by ore-minerals. Anhedral areas are chalcopyrite and bornite. Subhedral crystals within the hypersthene are magnetite.

x 11

FIG. 59

A higher magnification of the plagioclase in the left side of fig. 58. The ore-minerals are "eating in" along albite twinning lamellae.

x 23

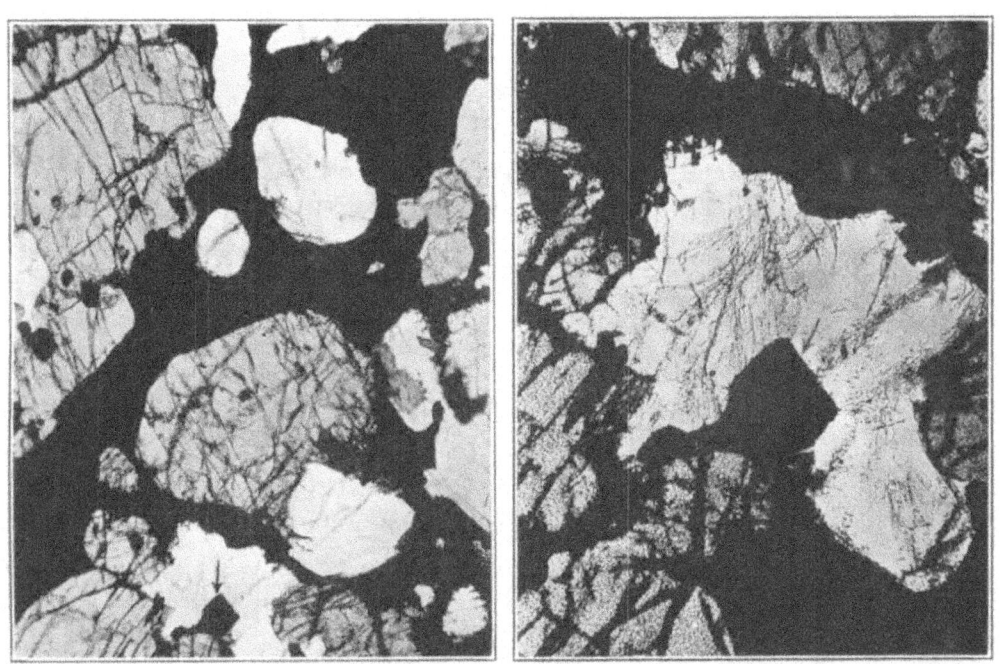

Fig. 56   Fig. 57

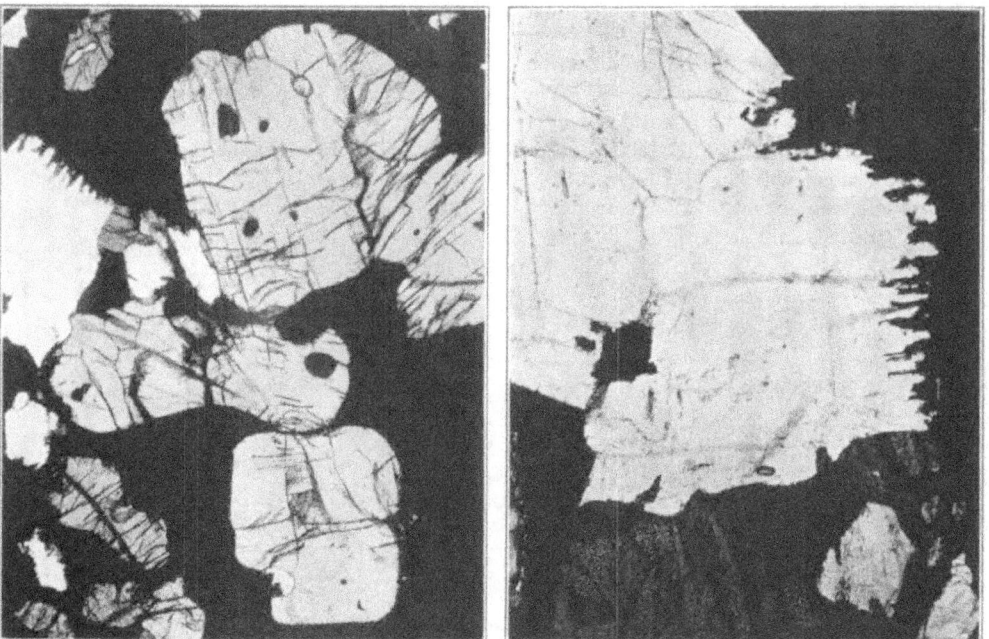

Fig. 58   Fig. 59

PLATE XV

PHOTOMICROGRAPHS OF A POLISHED SECTION OF ORE-BEARING NORITE,
TWEEFONTEIN MINE, OOKIEP, SOUTH AFRICA

FIG. 60

The ore-minerals shown in the slide are bornite [b], chalcopyrite [cp], magnetite [m], lined with ilmenite. (Hematite is shown in fig. 61.) Silicates are plagioclase [pl], hypersthene [hy], biotite [bi], and apatie [ap]. Note especially magnetite cutting hypersthene in the upper portion of the photograph, and bornite and chalcopyrite penetrating biotite on the left.

x 9

FIG. 61

Photograph from the same polished section as shown in fig. 60, showing hematite [h], and magnetite [m] intergrown with ilmenite lamellae.

x 18

FIG. 62

Photograph from the same section as shown in figs. 60 and 61. Chalcopyrite $[cp]_2$, and bornite [b] cut by a veinlet of anthopyllite [black needles], along which is developed a second generation of chalcopyrite (shown faintly at $[cp]_2$).

x 150

FIG. 63

A higher magnification of the veinlet shown in fig. 62. Bornite [gray], anthophyllite [black], and chalcopyrite of the second generation [white].

x 920

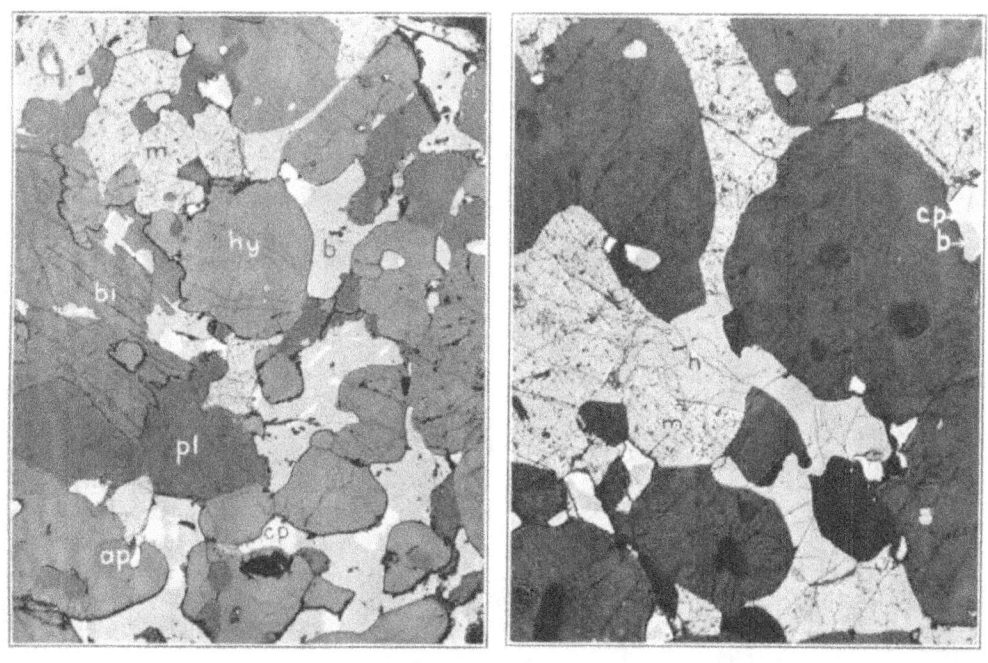

Fig. 60   Fig. 61

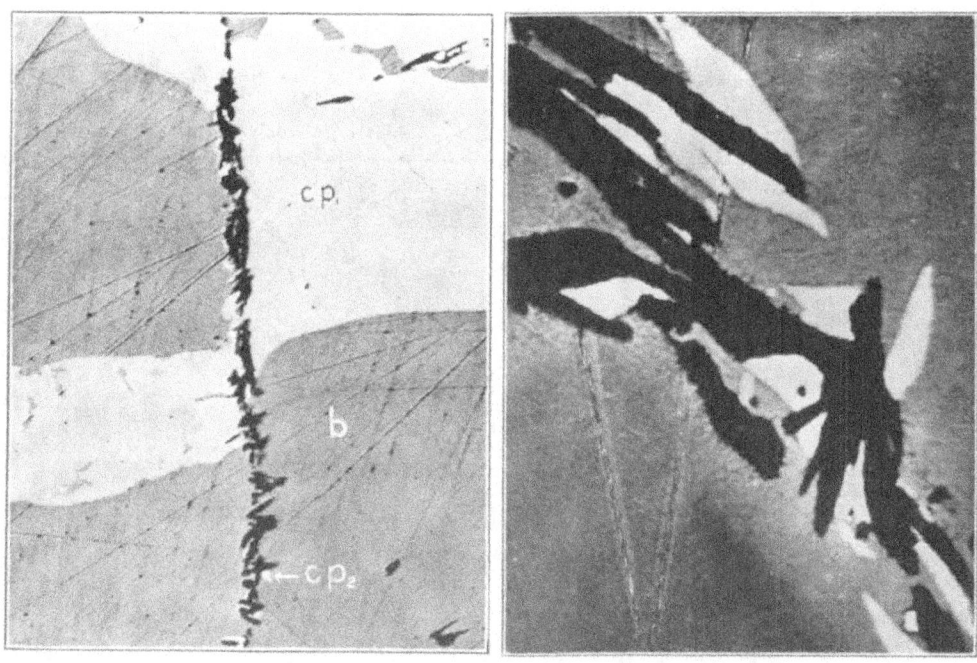

Fig. 62   Fig. 63

PLATE XVI

PHOTOMICROGRAPHS OF SECTIONS OF ORE-BEARING MICA DIORITE, OOKIEP EAST MINE, OOKIEP, SOUTH AFRICA

FIG. 64

Polished section showing pyrrhotite [p] residual in chalcopyrite [cp] and sulfid veinlets in the silicates.

x 10

FIG. 65

Thin section. The ore-minerals [black] are pyrrhotite and chalcopyrite. The primary silicate is plagioclase [gray]. The secondary silicates cut the ore-minerals and plagioclase as veinlets in the lower portion of photograph, and are distributed as specks in the sulfids. The alteration is subsequent to ore-formation.

x 18

FIG. 66

From the same thin section as fig. 65. The veinlets of sulfids are not a later generation of ore, nor are they due to rearrangements. This fact is shown in fig. 69.

x 45

FIG. 67

A group of subhedral magnetite crystals [m], a few of which are partially replaced by pyrrhotite [p] [at the bottom of the photograph].

x 51

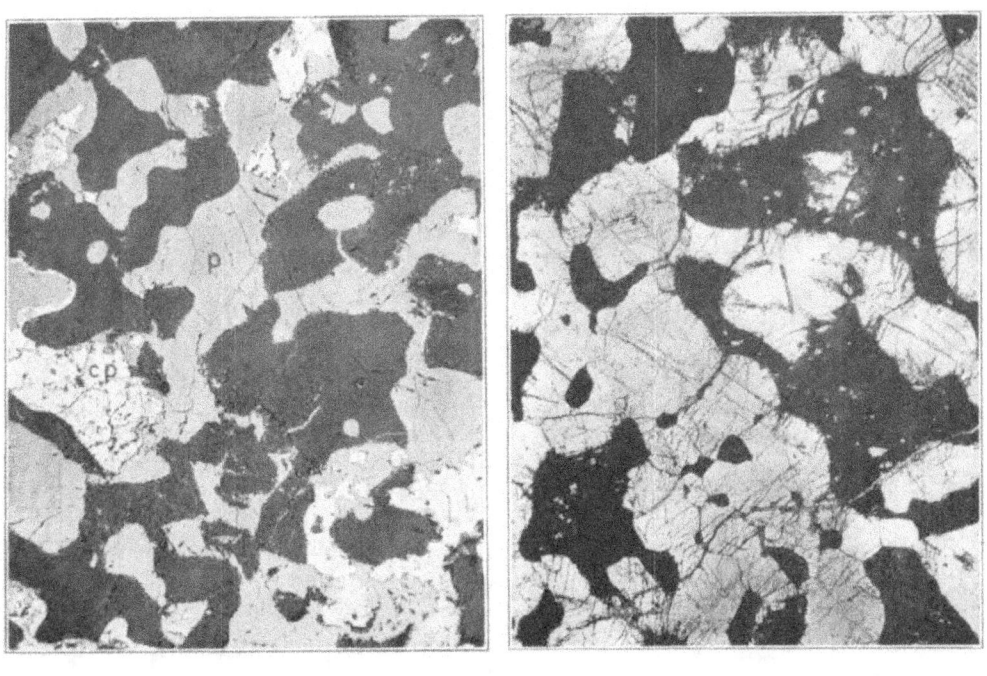

Fig. 64   Fig. 65

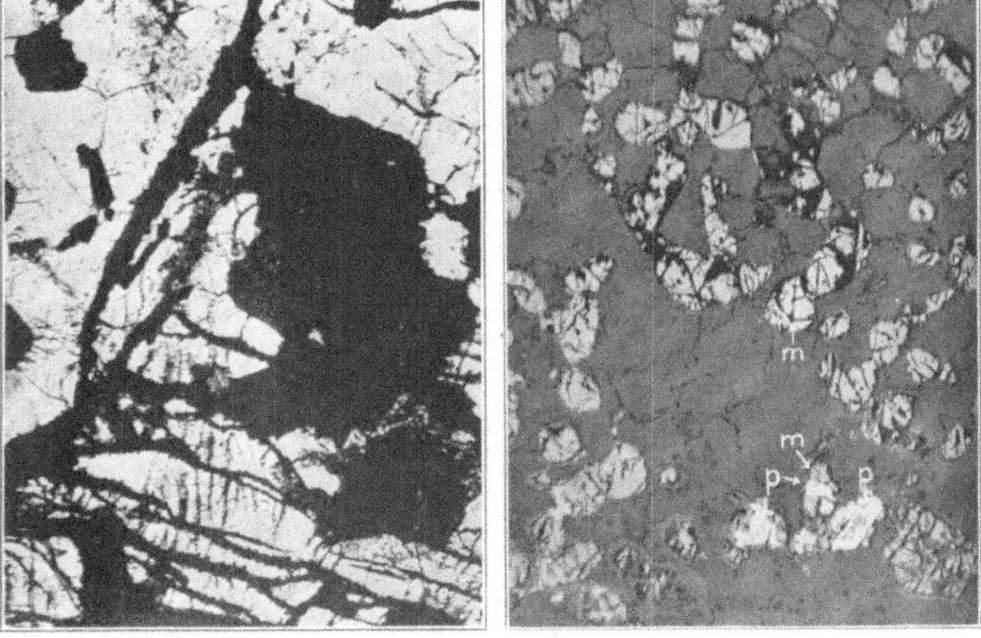

Fig. 66   Fig. 67

PLATE XVII

PHOTOMICROGRAPHS OF POLISHED SECTIONS OF ORE-BEARING MICA DIORITE,
OOKIEP EAST MINE, OOKIEP, SOUTH AFRICA

FIG. 68

Pyrrhotite [*p*] and chalcopyrite [*cp*] cutting the feldspars in sharp veinlets and penetrating the cleavage planes of the biotite [top of photograph].

x 10

FIG. 69

The chalcopyrite veinlet on the right is cut sharply by chloritic alteration products, and therefore the veinlets are not connected with the rock alteration, but antedate the latter. On the left a biotite crystal [*bi*] is penetrated by chalcopyrite along cleavage planes.

x 153

FIG. 70

A biotite-rich segregation in the diorite, with the sulfids, chalcopyrite [*cp*] and pyrrhotite [*p*], penetrating the cleavages of the biotite.

x 11

FIG. 71

Brush-like pentlandite (?) [*pn*]$_2$ replacing pyrrhotite [*p*] and chalcopyrite [*cp*]. It is probably of a late generation, like that shown in figs. 18, 35, 36, 42 and 43.

x 620

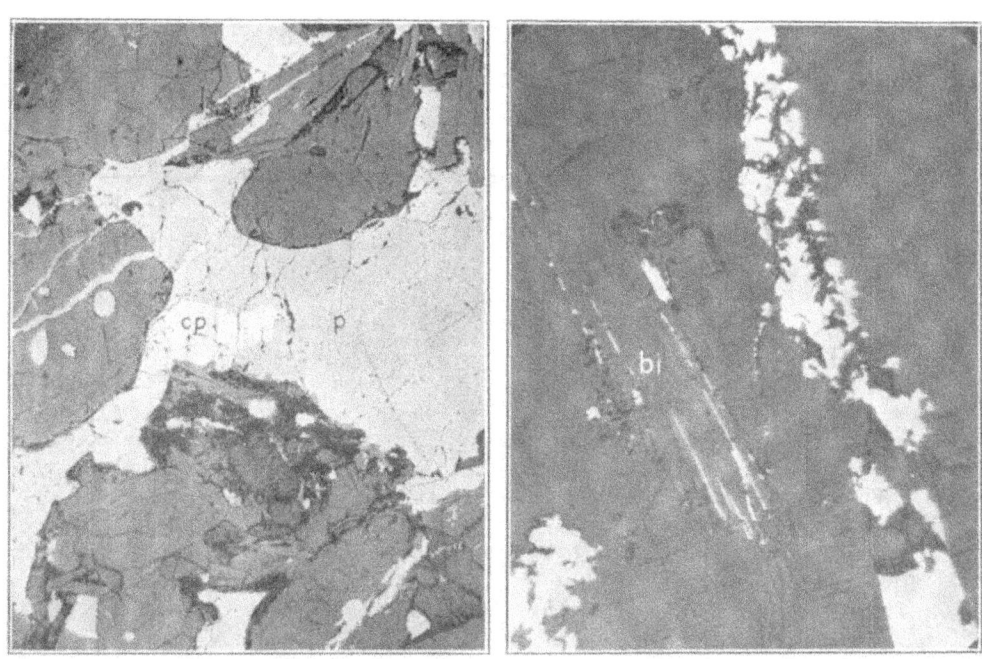

Fig. 68   Fig. 69

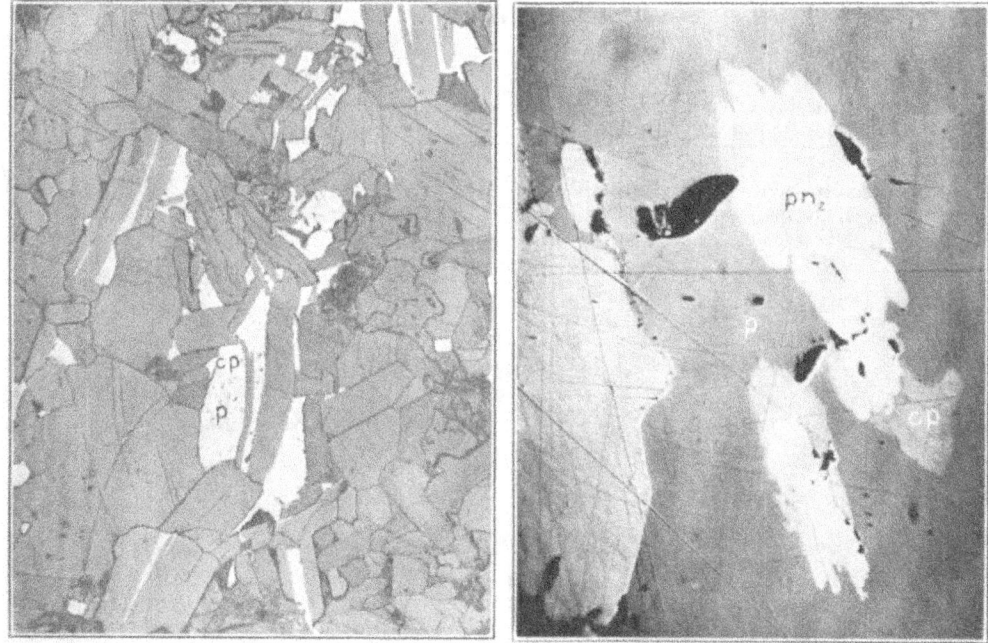

Fig. 70   Fig. 71

PLATE XVIII

Fig. 72

Unaltered norite-diorite showing the general relations of the ore-minerals [black] to the silicates (hypersthene [dark gray] and feldspar [light gray]). The ore-minerals are chiefly magnetite and hematite. The magnetite shows all gradations in form from euhedral to anhedral outlines.

x 22

Fig. 73

A higher magnification of a spot in the upper right-hand corner of fig. 72. Euhedral magnetite occurs within the silicates, anhedral along the boundaries. One area [indicated by arrow] shows an euhedral termination penetrating a silicate crystal, and the remaining portion is anhedral.

x 143

Fig. 74

A higher magnification of another spot in the upper right corner of fig. 72. Hook-shaped anhedra of magnetite and hematite [black] penetrating and replacing the silicates, especially biotite [bi].

x 195

Fig. 75

Hornblende [ho] rim surrounding pyroxenes [px] (diopside and hypersthene). This phenomenon is often observed in magmatic deposits.

x 40

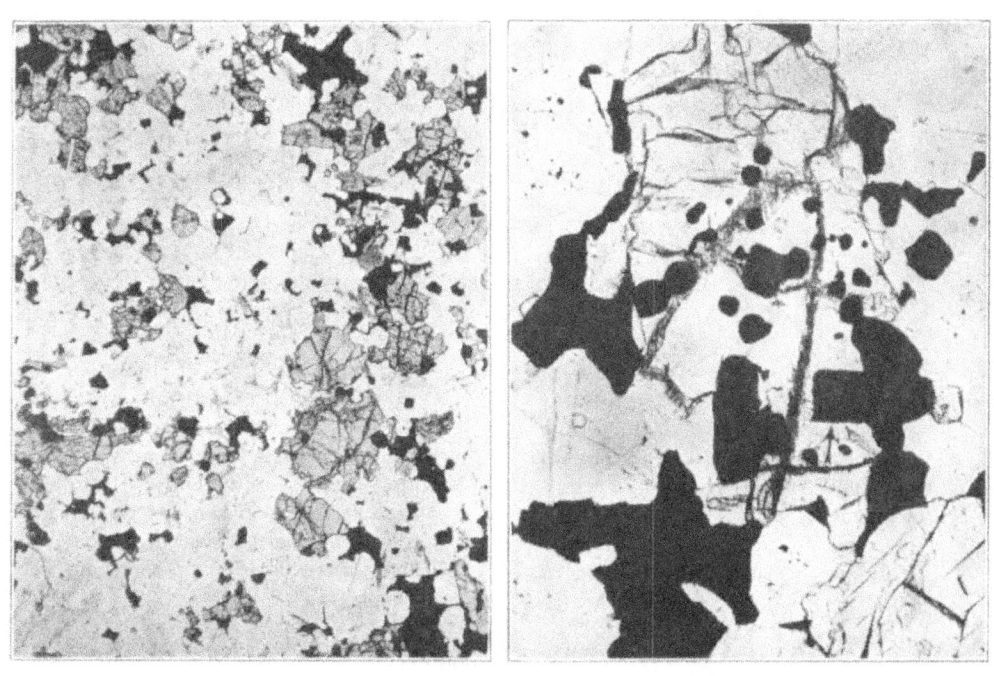

Fig. 72   Fig. 73

Fig. 74   Fig. 75

PLATE XIX

PHOTOMICROGRAPHS OF SECTIONS OF ROCKS AND ORES FROM THE ENGELS
MINE, PLUMAS COUNTY, CALIFORNIA

FIG. 76

Thin section of norite-diorite
showing diopside [di] with rim of
hornblende [ho]. Euhedral crystals
of magnetite [black] within the diopside, and anhedral crystals of magnetite and hematite [black] along its
borders.

x 63

FIG. 77

Thin section of norite-diorite
showing silicates [light gray and
white] and ore-minerals [black],
magnetite and hematite, cut by veinlets of chlorite (shown by arrow),
which proves that the alteration is
post-mineral.

x 36

FIG. 78

Thin section of ore-bearing grano-
diorite containing tourmaline [dark
gray], sericitized feldspars [light
gray], and ore-minerals [black],
bornite and chalcocite. The feldspars and tourmaline are replaced by
the ore-minerals and they in turn by
sericite laths. (One is indicated by
arrow in the lower right-hand corner.) This section shows local development of sericite. Most specimens, however, are not affected by
sericitization.

x 23

FIG. 79

Polished section of the ore shown
in fig. 78. Note the irregular hook-
shaped bornite [b] with narrow rim
of supergene chalcocite $[cc]_2$. The
bornite surrounds and replaces the
silicates.

x 184

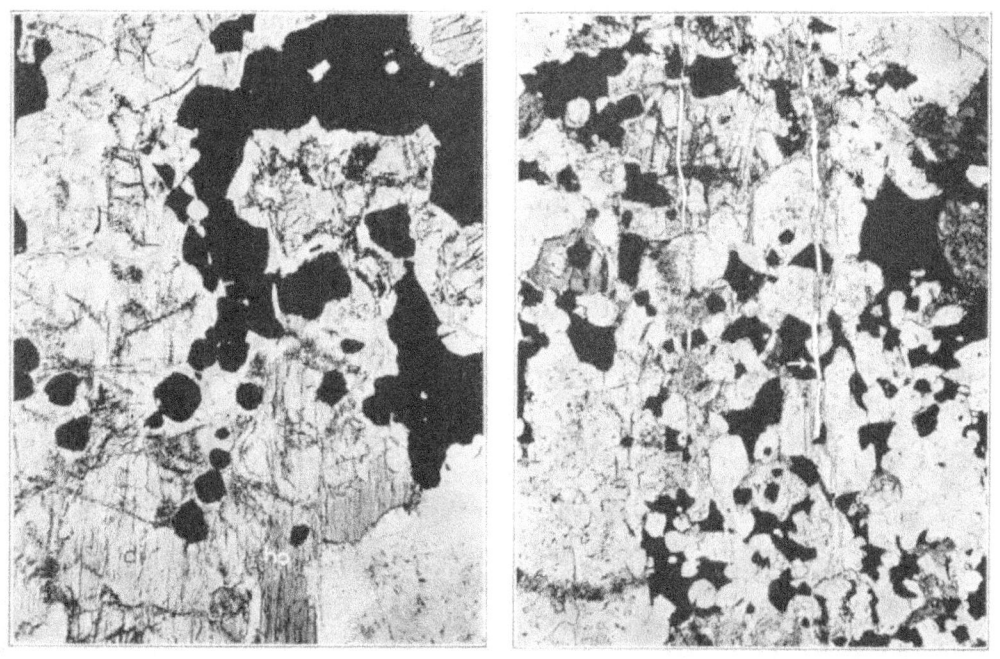

Fig. 76

Fig. 77

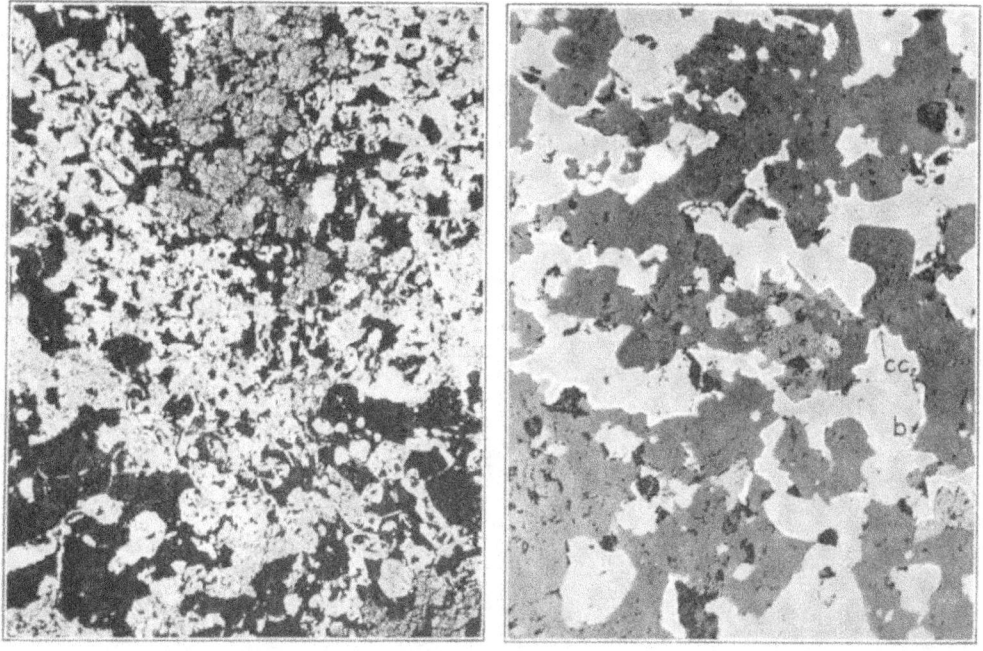

Fig. 78

Fig. 79

PLATE XX

PHOTOMICROGRAPHS OF POLISHED SECTIONS OF ORES FROM THE ENGELS
MINE, PLUMAS COUNTY, CALIFORNIA, SHOWING LOCAL
REARRANGEMENT AND COPPER ENRICHMENT SUB-
SEQUENT TO THE MAGMATIC STAGE

FIG. 80

The photograph shows the following post-magmatic alterations of bornite [b]: First the development of the so-called graphic intergrowth of bornite [b] and chalcocite $[cc]_1$, and later the development of rims of chalcocite $[cc]_2$ around the margins of the bornite areas. The first generation of chalcocite is probably hypogene, the second generation probably supergene.

x 125

FIG. 81

Anhedron of bornite [b] altering to covellite [cv] and chalcopyrite $[cp]_2$ of the second generation. The latter develops along crystallographic directions of the bornite. This alteration is probably supergene.

x 542

FIG. 82

An area of bornite [b] has been replaced by covellite [cv] and chalcocite $[cc]_1$, and later all of these have been penetrated by chlorite [ch] laths. This alteration is in part along crystallographic directions of the bornite.

x 142

FIG. 83

An area of bornite [b] has been replaced by chlorite [ch] laths and by quartz [q] veinlets. Chalcocite $[cc]_2$ of the second generation has developed along the margin of the bornite, along the chlorite-bornite contacts, and along veinlets.

x 142

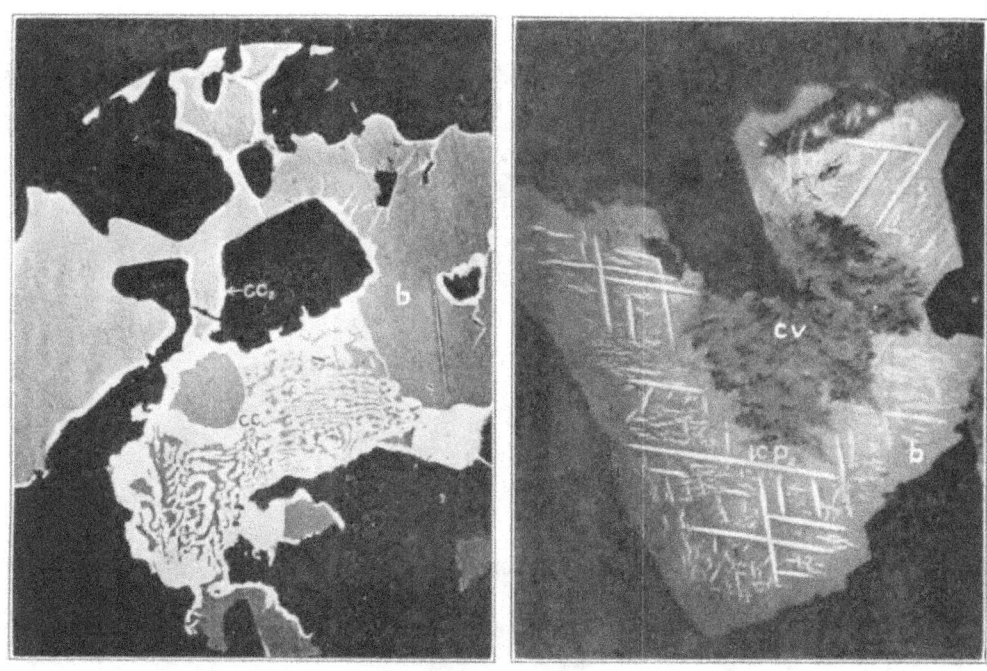

Fig. 80   Fig. 81

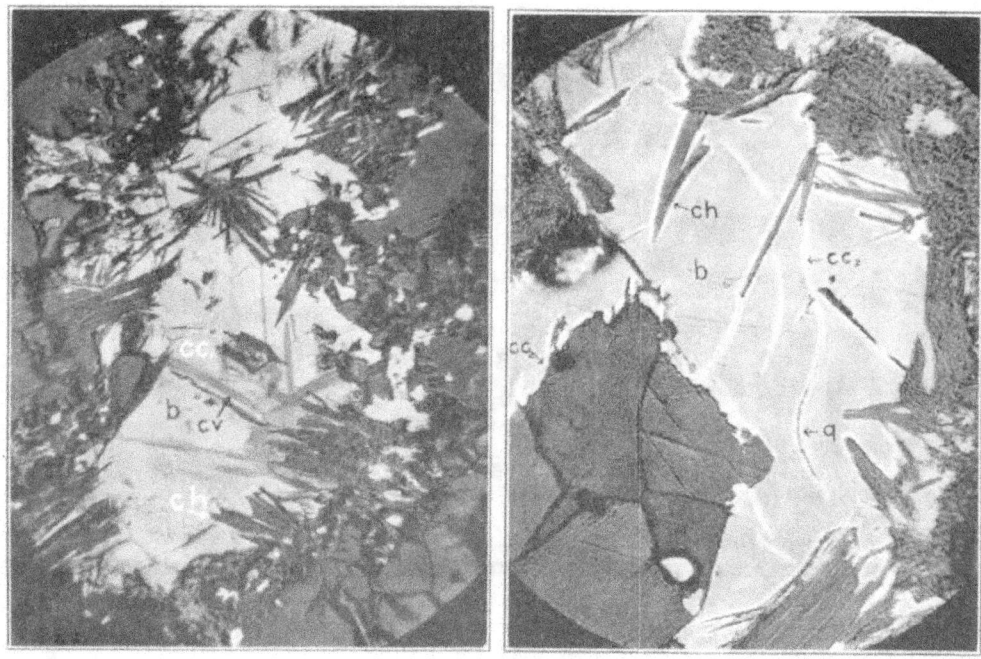

Fig. 82   Fig. 83

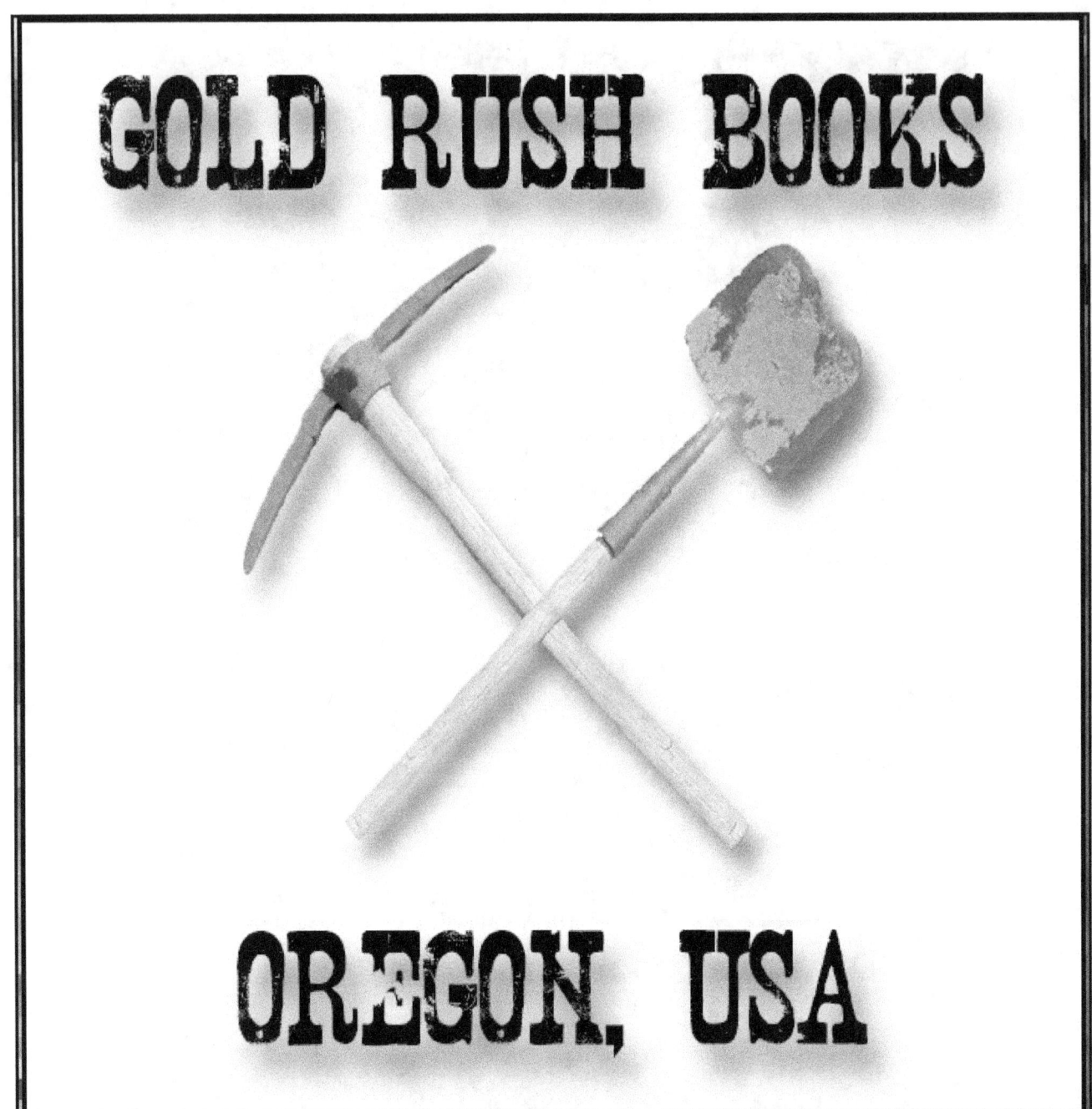

GOLD RUSH BOOKS

OREGON, USA

www.GoldMiningBooks.com

Books On Mining

Visit: www.goldminingbooks.com to order your copies or ask your favorite book seller to offer them.

Mining Books by Kerby Jackson

Gold Dust: Stories From Oregon's Mining Years - Oregon mining historian and prospector, Kerby Jackson, brings you a treasure trove of seventeen stories on Southern Oregon's rich history of gold prospecting, the prospectors and their discoveries, and the breathtaking areas they settled in and made homes. 5" X 8", 98 ppgs. **Retail Price: $11.99**

The Golden Trail: More Stories From Oregon's Mining Years - In his follow-up to "Gold Dust: Stories of Oregon's Mining Years", this time around, Jackson brings us twelve tales from Oregon's Gold Rush, including the story about the first gold strike on Canyon Creek in Grant County, about the old timers who found gold by the pail full at the Victor Mine near Galice, how Iradel Bray discovered a rich ledge of gold on the Coquille River during the height of the Rogue River War, a tale of two elderly miners on the hunt for a lost mine in the Cascade Mountains, details about the discovery of the famous Armstrong Nugget and others. 5" X 8", 70 ppgs. **Retail Price: $10.99**

Oregon Mining Books

Geology and Mineral Resources of Josephine County, Oregon - Unavailable since the 1970's, this important publication was originally compiled by the Oregon Department of Geology and Mineral Industries and includes important details on the economic geology and mineral resources of this important mining area in South Western Oregon. Included are notes on the history, geology and development of important mines, as well as insights into the mining of gold, copper, nickel, limestone, chromium and other minerals found in large quantities in Josephine County, Oregon. 8.5" X 11", 54 ppgs. **Retail Price: $9.99**

Mines and Prospects of the Mount Reuben Mining District - Unavailable since 1947, this important publication was originally compiled by geologist Elton Youngberg of the Oregon Department of Geology and Mineral Industries and includes detailed descriptions, histories and the geology of the Mount Reuben Mining District in Josephine County, Oregon. Included are notes on the history, geology, development and assay statistics, as well as underground maps of all the major mines and prospects in the vicinity of this much neglected mining district. 8.5" X 11", 48 ppgs. **Retail Price: $9.99**

The Granite Mining District - Notes on the history, geology and development of important mines in the well known Granite Mining District which is located in Grant County, Oregon. Some of the mines discussed include the Ajax, Blue Ribbon, Buffalo, Continental, Cougar-Independence, Magnolia, New York, Standard and the Tillicum. Also included are many rare maps pertaining to the mines in the area. 8.5" X 11", 48 ppgs. **Retail Price: $9.99**

Ore Deposits of the Takilma and Waldo Mining Districts of Josephine County, Oregon - The Waldo and Takilma mining districts are most notable for the fact that the earliest large scale mining of placer gold and copper in Oregon took place in these two areas. Included are details about some of the earliest large gold mines in the state such as the Llano de Oro, High Gravel, Cameron, Platerica, Deep Gravel and others, as well as copper mines such as the famous Queen of Bronze mine, the Waldo, Lily and Cowboy mines. This volume also includes six maps and 20 original illustrations. 8.5" X 11", 74 ppgs. **Retail Price: $9.99**

Metal Mines of Douglas, Coos and Curry Counties, Oregon - Oregon mining historian Kerby Jackson introduces us to a classic work on Oregon's mining history in this important re-issue of Bulletin 14C Volume 1, otherwise known as the Douglas, Coos & Curry Counties, Oregon Metal Mines Handbook. Unavailable since 1940, this important publication was originally compiled by the Oregon Department of Geology and Mineral Industries includes detailed descriptions, histories and the geology of over 250 metallic mineral mines and prospects in this rugged area of South West Oregon. 8.5" X 11", 158 ppgs. **Retail Price: $19.99**

Metal Mines of Jackson County, Oregon - Unavailable since 1943, this important publication was originally compiled by the Oregon Department of Geology and Mineral Industries includes detailed descriptions, histories and the geology of over 450 metallic mineral mines and prospects in Jackson County, Oregon. Included are such famous gold mining areas as Gold Hill, Jacksonville, Sterling and the Upper Applegate. **8.5" X 11", 220 ppgs. Retail Price: $24.99**

Metal Mines of Josephine County, Oregon - Oregon mining historian Kerby Jackson introduces us to a classic work on Oregon's mining history in this important re-issue of Bulletin 14C, otherwise known as the Josephine County, Oregon Metal Mines Handbook. Unavailable since 1952, this important publication was originally compiled by the Oregon Department of Geology and Mineral Industries includes detailed descriptions, histories and the geology of over 500 metallic mineral mines and prospects in Josephine County, Oregon. **8.5" X 11", 250 ppgs. Retail Price: $24.99**

Metal Mines of North East Oregon - Oregon mining historian Kerby Jackson introduces us to a classic work on Oregon's mining history in this important re-issue of Bulletin 14A and 14B, otherwise known as the North East Oregon Metal Mines Handbook. Unavailable since 1941, this important publication was originally compiled by the Oregon Department of Geology and Mineral Industries and includes detailed descriptions, histories and the geology of over 750 metallic mineral mines and prospects in North Eastern Oregon. **8.5" X 11", 310 ppgs. Retail Price: $29.99**

Metal Mines of North West Oregon - Oregon mining historian Kerby Jackson introduces us to a classic work on Oregon's mining history in this important re-issue of Bulletin 14D, otherwise known as the North West Oregon Metal Mines Handbook. Unavailable since 1951, this important publication was originally compiled by the Oregon Department of Geology and Mineral Industries and includes detailed descriptions, histories and the geology of over 250 metallic mineral mines and prospects in North Western Oregon. **8.5" X 11", 182 ppgs. Retail Price: $19.99**

Mines and Prospects of Oregon - Mining historian Kerby Jackson introduces us to a classic mining work by the Oregon Bureau of Mines in this important re-issue of The Handbook of Mines and Prospects of Oregon. Unavailable since 1916, this publication includes important insights into hundreds of gold, silver, copper, coal, limestone and other mines that operated in the State of Oregon around the turn of the 19th Century. Included are not only geological details on early mines throughout Oregon, but also insights into their history, production, locations and in some cases, also included are rare maps of their underground workings. **8.5" X 11", 314 ppgs. Retail Price: $24.99**

Lode Gold of the Klamath Mountains of Northern California and South West Oregon
(See California Mining Books)

Mineral Resources of South West Oregon - Unavailable since 1914, this publication includes important insights into dozens of mines that once operated in South West Oregon, including the famous gold fields of Josephine and Jackson Counties, as well as the Coal Mines of Coos County. Included are not only geological details on early mines throughout South West Oregon, but also insights into their history, production and locations. **8.5" X 11", 154 ppgs. Retail Price: $11.99**

Chromite Mining in The Klamath Mountains of California and Oregon
(See California Mining Books)

Southern Oregon Mineral Wealth - Unavailable since 1904, this rare publication provides a unique snapshot into the mines that were operating in the area at the time. Included are not only geological details on early mines throughout South West Oregon, but also insights into their history, production and locations. Some of the mining areas include Grave Creek, Greenback, Wolf Creek, Jump Off Joe Creek, Granite Hill, Galice, Mount Reuben, Gold Hill, Galls Creek, Kane Creek, Sardine Creek, Birdseye Creek, Evans Creek, Foots Creek, Jacksonville, Ashland, the Applegate River, Waldo, Kerby and the Illinois River, Althouse and Sucker Creek, as well as insights into local copper mining and other topics. **8.5" X 11", 64 ppgs. Retail Price: $8.99**

Geology and Ore Deposits of the Takilma and Waldo Mining Districts - Unavailable since the 1933, this publication was originally compiled by the United States Geological Survey and includes details on gold and copper mining in the Takilma and Waldo Districts of Josephine County, Oregon. The Waldo and Takilma mining districts are most notable for the fact that the earliest large scale mining of placer gold and copper in Oregon took place in these two areas. Included in this report are details about some of the earliest large gold mines in the state such as the Llano de Oro, High Gravel, Cameron, Platerica, Deep Gravel and others, as well as copper mines such as the famous Queen of Bronze mine, the Waldo, Lily and Cowboy mines. In addition to geological examinations, insights are also provided into the production, day to day operations and early histories of these mines, as well as calculations of known mineral reserves in the area. This volume also includes six maps and 20 original illustrations. **8.5" X 11", 74 ppgs. Retail Price: $9.99**

Gold Mines of Oregon - Oregon mining historian Kerby Jackson introduces us to a classic work on Oregon's mining history in this important re-issue of Bulletin 61, otherwise known as "Gold and Silver In Oregon". Unavailable since 1968, this important publication was originally compiled by geologists Howard C. Brooks and Len Ramp of the Oregon Department of Geology and Mineral Industries and includes detailed descriptions, histories and the geology of over 450 gold mines Oregon. Included are notes on the history, geology and gold production statistics of all the major mining areas in Oregon including the Klamath Mountains, the Blue Mountains and the North Cascades. While gold is where you find it, as every miner knows, the path to success is to prospect for gold where it was previously found. **8.5" X 11", 344 ppgs. Retail Price: $24.99**

Mines and Mineral Resources of Curry County Oregon - Originally published in 1916, this important publication on Oregon Mining has not been available for nearly a century. Included are rare insights into the history, production and locations of dozens of gold mines in Curry County, Oregon, as well as detailed information on important Oregon mining districts in that area such as those at Agness, Bald Face Creek, Mule Creek, Boulder Creek, China Diggings, Collier Creek, Elk River, Gold Beach, Rock Creek, Sixes River and elsewhere. Particular attention is especially paid to the famous beach gold deposits of this portion of the Oregon Coast. **8.5" X 11", 140 ppgs. Retail Price: $11.99**

Chromite Mining in South West Oregon - Originally published in 1961, this important publication on Oregon Mining has not been available for nearly a century. Included are rare insights into the history, production and locations of nearly 300 chromite mines in South Western Oregon. **8.5" X 11", 184 ppgs. Retail Price: $14.99**

Mineral Resources of Douglas County Oregon - Originally published in 1972, this important publication on Oregon Mining has not been available for nearly forty years. Included are rare insights into the geology, history, production and locations of numerous gold mines and other mining properties in Douglas County, Oregon. **8.5" X 11", 124 ppgs. Retail Price: $11.99**

Mineral Resources of Coos County Oregon - Originally published in 1972, this important publication on Oregon Mining has not been available for nearly forty years. Included are rare insights into the geology, history, production and locations of numerous gold mines and other mining properties in Coos County, Oregon. **8.5" X 11", 100 ppgs. Retail Price: $11.99**

Mineral Resources of Lane County Oregon - Originally published in 1938, this important publication on Oregon Mining has not been available for nearly seventy five years. Included are extremely rare insights into the geology and mines of Lane County, Oregon, in particular in the Bohemia, Blue River, Oakridge, Black Butte and Winberry Mining Districts. **8.5" X 11", 82 ppgs. Retail Price: $9.99**

Mineral Resources of the Upper Chetco River of Oregon: Including the Kalmiopsis Wilderness - Originally published in 1975, this important publication on Oregon Mining has not been available for nearly forty years. Withdrawn under the 1872 Mining Act since 1984, real insight into the minerals resources and mines of the Upper Chetco River has long been unavailable due to the remoteness of the area. Despite this, the decades of battle between property owners and environmental extremists over the last private mining inholding in the area has continued to pique the interest of those interested in mining and other forms of natural resource use. Gold mining began in the area in the 1850's and has a rich history in this geographic area, even if the facts surrounding it are little known. Included are twenty two rare photographs, as well as insights into the Becca and Morning Mine, the Emmly Mine (also known as Emily Camp), the Frazier Mine, the Golden Dream or Higgins Mine, Hustis Mine, Peck Mine and others. **8.5" X 11", 64 ppgs. Retail Price: $8.99**

Gold Dredging in Oregon - Originally published in 1939, this important publication on Oregon Mining has not been available for nearly seventy five years. Included are extremely rare insights into the history and day to day operations of the dragline and bucketline gold dredges that once worked the placer gold fields of South West and North East Oregon in decades gone by. Also included are details into the areas that were worked by gold dredges in Josephine, Jackson, Baker and Grant counties, as well as the economic factors that impacted this mining method. This volume also offers a unique look into the values of river bottom land in relation to both farming and mining, in how farm lands were mined, re-soiled and reclamated after the dredges worked them. Featured are hard to find maps of the gold dredge fields, as well as rare photographs from a bygone era. **8.5" X 11", 86 ppgs. Retail Price: $8.99**

Quick Silver Mining in Oregon - Originally published in 1963, this important publication on Oregon Mining has not been available for over fifty years. This publication includes details into the history and production of Elemental Mercury or Quicksilver in the State of Oregon. **8.5" X 11", 238 ppgs. Retail Price: $15.99**

Mines of the Greenhorn Mining District of Grant County Oregon - Originally published in 1948, this important publication on Oregon Mining has not been available for over sixty five years. In this publication are rare insights into the mines of the famous Greenhorn Mining District of Grant County, Oregon, especially the famous Morning Mine. Also included are details on the Tempest, Tiger, Bi-Metallic, Windsor, Psyche, Big Johnny, Snow Creek, Banzette and Paramount Mines, as well as prospects in the vicinities in the famous mining areas of Mormon Basin, Vinegar Basin and Desolation Creek. Included are hard to find mine maps and dozens of rare photographs from the bygone era of Grant County's rich mining history. **8.5" X 11", 72 ppgs. Retail Price: $9.99**

Geology of the Wallowa Mountains of Oregon: Part I (Volume 1) - Originally published in 1938, this important publication on Oregon Mining has not been available for nearly seventy five years. Included are details on the geology of this unique portion of North Eastern Oregon. This is the first part of a two book series on the area. Accompanying the text are rare photographs and historic maps.**8.5″ X 11″, 92 ppgs. Retail Price: $9.99**

Geology of the Wallowa Mountains of Oregon: Part II (Volume 2) - Originally published in 1938, this important publication on Oregon Mining has not been available for nearly seventy five years. Included are details on the geology of this unique portion of North Eastern Oregon. This is the first part of a two book series on the area. Accompanying the text are rare photographs and historic maps.**8.5″ X 11″, 94 ppgs. Retail Price: $9.99**

Field Identification of Minerals For Oregon Prospectors - Originally published in 1940, this important publication on Oregon Mining has not been available for nearly seventy five years. Included in this volume is an easy system for testing and identifying a wide range of minerals that might be found by prospectors, geologists and rockhounds in the State of Oregon, as well as in other locales. Topics include how to put together your own field testing kit and how to conduct rudimentary tests in the field. This volume is written in a clear and concise way to make it useful even for beginners. **8.5″ X 11″, 158 ppgs. Retail Price: $14.99**

The Bohemia Mining District of Oregon - Originally published in 1900, this important publication on Oregon Mining has not been available for over a century. Included in this volume are important insights into the famous Bohemia Mining District of Oregon, including the histories and locations of important gold mines in the area such as the Ophir Mine, Clarence, Acturas, Peek-a-boo, White Swan, Combination Mine, the Musick Mine, The California, White Ghost, The Mystery, Wall Street, Vesuvius, Story, Lizzie Bullock, Delta, Elsie Dora, Golden Slipper, Broadway, Champion Mine, Knott, Noonday, Helena, White Wings, Riverside and others. Also included are notes on the nearby Blue River Mining District. **8.5″ X 11″, 58 ppgs. Retail Price: $9.99**

The Gold Fields of Eastern Oregon - Unavailable since 1900, this publication was originally compiled by the Baker City Chamber of Commerce Offering important insights into the gold mining history of Eastern Oregon, "The Gold Fields of Eastern Oregon" sheds a rare light on many of the gold mines that were operating at the turn of the 19th Century in Baker County and Grant County in North Eastern Oregon. Some of the areas featured include the Cable Cove District, Baisely-Elhorn, Granite, Red Boy, Bonanza, Susanville, Sparta, Virtue, Vaughn, Sumpter, Burnt River, Rye Valley and other mining districts. Included is basic information on not only many gold mines that are well known to those interested in Eastern Oregon mining history, but also many mines and prospects which have been mostly lost to the passage of time. Accompanying are numerous rare photos **8.5″ X 11″, 78 ppgs. Retail Price: $10.99**

Gold Mining in Eastern Oregon - Originally published in 1938, this important publication on Oregon Mining has not been available for over a century. Included in this volume are important insights into the famous mining districts of Eastern Oregon during the late 1930's. Particular attention is given to those gold mines with milling and concentrating facilities in the Greenhorn, Red Boy, Alamo, Bonanza, Granite, Cable Cove, Cracker Creek, Virtue, Keating, Medical Springs, Sanger, Sparta, Chicken Creek, Mormon Basin, Connor Creek, Cornucopia and the Bull Run Mining Districts. Some of the mines featured include the Ben Harrison, North Pole-Columbia, Highland Maxwell, Baisley-Elkhorn, White Swan, Balm Creek, Twin Baby, Gem of Sparta, New Deal, Gleason, Gifford-Johnson, Cornucopia, Record, Bull Run, Orion and others. Of particular interest are the mill flow sheets and descriptions of milling operations of these mines. **8.5″ X 11″, 68 ppgs. Retail Price: $8.99**

The Gold Belt of the Blue Mountains of Oregon - Originally published in 1901, this important publication on Oregon Mining has not been available for over a century. Included in this volume are rare insights into the gold deposits of the Blue Mountains of North East Oregon, including the history of their early discovery and early production. Extensive details are offered on this important mining area's mineralogy and economic geology, as well as insights into nearby gold placers, silver deposits and copper deposits. Featured are the Elkhorn and Rock Creek mining districts, the Pocahontas district, Auburn and Minersville districts, Sumpter and Cracker Creek, Cable Cove, the Camp Carson district, Granite, Alamo, Greenhorn, Robinsonville, the Upper Burnt River Valley and Bonanza districts, Susanville, Quartzburg, Canyon Creek, Virtue, the Copper Butte district, the North Powder River, Sparta, Eagle Creek, Cornucopia, Pine Creek, Lower Powder River, the Upper Snake River Canyon, Rye Valley, Lower Burnt River Valley, Mormon Basin, the Malheur and Clarks Creek districts, Sutton Creek and others. Of particular interest are important details on numerous gold mines and prospects in these mining districts, including their locations, histories, geology and other important information, as well as information on silver, copper and fire opal deposits. **8.5″ X 11″, 250 ppgs. Retail Price: $24.99**

<u>Mining in the Cascades Range of Oregon</u> - Originally published in 1938, this important publication on Oregon Mining has not been available for over seventy five years. Included in this volume are rare insights into the gold mines and other types of metal mines in the Cascades Mountain Range of Oregon. Some of the important mining areas covered include the famous Bohemia Mining District, the North Santiam Mining District, Quartzville Mining District, Blue River Mining District, Fall Creek Mining District, Oakridge District, Zinc District, Buzzard-Al Sarena District, Grand Cove, Climax District and Barron Mining District. Of particular interest are important details on over 100 mines and prospects in these mining districts, including their locations, histories, geology and other important information. **8.5" X 11", 170 ppgs. Retail Price: $14.99**

Idaho Mining Books

<u>Gold in Idaho</u> - Unavailable since the 1940's, this publication was originally compiled by the Idaho Bureau of Mines and includes details on gold mining in Idaho. Included is not only raw data on gold production in Idaho, but also valuable insight into where gold may be found in Idaho, as well as practical information on the gold bearing rocks and other geological features that will assist those looking for placer and lode gold in the State of Idaho. This volume also includes thirteen gold maps that greatly enhance the practical usability of the information contained in this small book detailing where to find gold in Idaho. **8.5" X 11", 72 ppgs. Retail Price: $9.99**

<u>Geology of the Couer D'Alene Mining District of Idaho</u> - Unavailable since 1961, this publication was originally compiled by the Idaho Bureau of Mines and Geology and includes details on the mining of gold, silver and other minerals in the famous Coeur D'Alene Mining District in Northern Idaho. Included are details on the early history of the Coeur D'Alene Mining District, local tectonic settings, ore deposit features, information on the mineral belts of the Osburn Fault, as well as detailed information on the famous Bunker Hill Mine, the Dayrock Mine, Galena Mine, Lucky Friday Mine and the infamous Sunshine Mine. This volume also includes sixteen hard to find maps. **8.5" X 11", 70 ppgs. Retail Price: $9.99**

<u>The Gold Camps and Silver Cities of Idaho</u> - Originally published in 1963, this important publication on Idaho Mining has not been available for nearly fifty years. Included are rare insights into the history of Idaho's Gold Rush, as well as the mad craze for silver in the Idaho Panhandle. Documented in fine detail are the early mining excitements at Boise Basin, at South Boise, in the Owyhees, at Deadwood, Long Valley, Stanley Basin and Robinson Bar, at Atlanta, on the famous Boise River, Volcano, Little Smokey, Banner, Boise Ridge, Hailey, Leesburg, Lemhi, Pearl, at South Mountain, Shoup and Ulysses, Yellow Jacket and Loon Creek. The story follows with the appearance of Chinese miners at the new mining camps on the Snake River, Black Pine, Yankee Fork, Bay Horse, Clayton, Heath, Seven Devils, Gibbonsville, Vienna and Sawtooth City. Also included are special sections on the Idaho Lead and Silver mines of the late 1800's, as well as the mining discoveries of the early 1900's that paved the way for Idaho's modern mining and mineral industry. Lavishly illustrated with rare historic photos, this volume provides a one of a kind documentary into Idaho's mining history that is sure to be enjoyed by not only modern miners and prospectors who still scour the hills in search of nature's treasures, but also those enjoy history and tromping through overgrown ghost towns and long abandoned mining camps. **8.5" X 11", 186 ppgs. Retail Price: $14.99**

<u>Ore Deposits and Mining in North Western Custer County Idaho</u> - Unavailable since 1913, this important publication was originally published by the Us Department of the Interior and has been unavailable for a century. Included are fine details on the geology, geography, gold placers and gold and silver bearing quartz veins of the mining region of North West Custer County, Idaho. Of particular interest is a rare look at the mines and prospects of the region, including those such as the Ramshorn Mine, SkyLark, Riverview, Excelsior, Beardsley, Pacific, Hoosier, Silver Brick, Forest Rose and dozens of others in the Bay Horse Mining District. Also covered are the mines of the Yankee Fork District such as the Lucky Boy, Badger, Black, Enterprise, Charles Dickens, Morrison, Golden Sunbeam, Montana, Golden Gate and others, as well as those in the Loon Mining District. **8.5" X 11", 126 ppgs. Retail Price: $12.99**

<u>Gold Rush To Idaho</u> - Unavailable since 1963, this important publication was originally published by the Idaho Bureau of Mines and has been unavailable for 50 years. "Gold Rush To Idaho" revisits the earliest years of the discovery of gold in Idaho Territory and introduces us to the conditions that the pioneer gold seekers met when they blazed a trail through the wilderness of Idaho's mountains and discovered the precious yellow metal at Oro Fino and Pierce. Subsequent rushes followed at places like Elk City, Newsome, Clearwater Station, Florence, Warrens and elsewhere. Of particular interest is a rare look at the hardships that the first miners in Idaho met with during their day to day existences and their attempts to bring law and order to their mining camps. **8.5" X 11", 88 ppgs. Retail Price: $9.99**

The Geology and Mines of Northern Idaho and North Western Montana - Unavailable since 1909, this important publication was originally published by the Us Department of the Interior and has been unavailable for a century. Included are fine details on the geology and geography of the mining regions of Northern Idaho and North Western Montana. Of particular interest is a rare look at the mines and prospects of the region, including those in the Pine Creek Mining District, Lake Pend Oreille district, Troy Mining District, Sylvanite District, Cabinet Mining District, Prospect Mining District and the Missoula Valley. Some of the mines featured include the Iron Mountain, Silver Butte, Snowshoe, Grouse Mountain Mine and others. **8.5" X 11", 142 ppgs. Retail Price: $12.99**

Utah Mining Books

Fluorite in Utah - Unavailable since 1954, this publication was originally compiled by the USGS, State of Utah and U.S. Atomic Energy Commission and details the mining of fluorspar, also known as fluorite in the State of Utah. Included are details on the geology and history of fluorspar (fluorite) mining in Utah, including details on where this unique gem mineral may be found in the State of Utah. **8.5" X 11", 60 ppgs. Retail Price: $8.99**

California Mining Books

The Tertiary Gravels of the Sierra Nevada of California - Mining historian Kerby Jackson introduces us to a classic mining work by Waldemar Lindgren in this important re-issue of The Tertiary Gravels of the Sierra Nevada of California. Unavailable since 1911, this publication includes details on the gold bearing ancient river channels of the famous Sierra Nevada region of California. **8.5" X 11", 282 ppgs. Retail Price: $19.99**

The Mother Lode Mining Region of California - Unavailable since 1900, this publication includes details on the gold mines of California's famous Mother Lode gold mining area. Included are details on the geology, history and important gold mines of the region, as well as insights into historic mining methods, mine timbering, mining machinery, mining bell signals and other details on how these mines operated. Also included are insights into the gold mines of the California Mother Lode that were in operation during the first sixty years of California's mining history. **8.5" X 11", 176 ppgs. Retail Price: $14.99**

Lode Gold of the Klamath Mountains of Northern California and South West Oregon - Unavailable since 1971, this publication was originally compiled by Preston E. Hotz and includes details on the lode mining districts of Oregon and California's Klamath Mountains. Included are details on the geology, history and important lode mines of the French Gulch, Deadwood, Whiskeytown, Shasta, Redding, Muletown, South Fork, Old Diggings, Dog Creek (Delta), Bully Choop (Indian Creek), Harrison Gulch, Hayfork, Minersville, Trinity Center, Canyon Creek, East Fork, New River, Denny, Liberty (Black Bear), Cecilville, Callahan, Yreka, Fort Jones and Happy Camp mining districts in California, as well as the Ashland, Rogue River, Applegate, Illinois River, Takilma, Greenback, Galice, Silver Peak, Myrtle Creek and Mule Creek districts of South Western Oregon. Also included are insights into the mineralization and other characteristics of this important mining region. **8.5" X 11", 100 ppgs. Retail Price: $10.99**

Mines and Mineral Resources of Shasta County, Siskiyou County, Trinity County: California - Unavailable since 1915, this publication was originally compiled by the California State Mining Bureau and includes details on the gold mines of this area of Northern California. Also included are insights into the mineralization and other characteristics of this important mining region, as well as the location of historic gold mines. **8.5" X 11", 204 ppgs. Retail Price: $19.99**

Geology of the Yreka Quadrangle, Siskiyou County, California - Unavailable since 1977, this publication was originally compiled by Preston E. Hotz and includes details on the geology of the Yreka Quadrangle of Siskiyou County, California. Also included are insights into the mineralization and other characteristics of this important mining region. **8.5" X 11", 78 ppgs. Retail Price: $7.99**

Mines of San Diego and Imperial Counties, California - Originally published in 1914, this important publication on California Mining has not been available for a century. This publication includes important information on the early gold mines of San Diego and Imperial County, which were some of the first gold fields mined in California by early Spanish and Mexican miners before the 49ers came on the scene. Included are not only details on early mining methods in the area, production statistics and geological information, but also the location of the early gold mines that helped make California "The Golden State". Also included are details on the mining of other minerals such as silver, lead, zinc, manganese, tungsten, vanadium, asbestos, barite, borax, cement, clay, dolomite, fluospar, gem stones, graphite, marble, salines, petroleum, stronium, talc and others. **8.5" X 11", 116 ppgs. Retail Price: $12.99**

Mines of Sierra County, California - Unavailable since 1920, this publication was originally compiled by the California State Mining Bureau and includes details on the gold mines of Sierra County, California. Also included are insights into the mineralization and other characteristics of this important mining region, as well as the location of historic gold mines. **8.5" X 11", 156 ppgs. Retail Price: $19.99**

Mines of Plumas County, California - Unavailable since 1918, this publication was originally compiled by the California State Mining Bureau and includes details on the gold mines of Plumas County, California. Also included are insights into the mineralization and other characteristics of this important mining region, as well as the location of historic gold mines. **8.5" X 11", 200 ppgs. Retail Price: $19.99**

Mines of El Dorado, Placer, Sacramento and Yuba Counties, California - Originally published in 1917, this important publication on California Mining has not been available for nearly a century. This publication includes important information on the early gold mines of El Dorado County, Placer County, Sacramento County and Yuba County, which were some of the first gold fields mined by the Forty-Niners during the California Gold Rush. Included are not only details on early mining methods in the area, production statistics and geological information, but also the location of the early gold mines that helped make California "The Golden State". Also included are insights into the early mining of chrome, copper and other minerals in this important mining area. **8.5" X 11", 204 ppgs. Retail Price: $19.99**

Mines of Los Angeles, Orange and Riverside Counties, California - Originally published in 1917, this important publication on California Mining has not been available for nearly a century. This publication includes important information on the early gold mines of Los Angeles County, Orange County and Riverside County, which were some of the first gold fields mined in California by early Spanish and Mexican miners before the 49ers came on the scene. Included are not only details on early mining methods in the area, production statistics and geological information, but also the location of the early gold mines that helped make California "The Golden State". **8.5" X 11", 146 ppgs. Retail Price: $12.99**

Mines of San Bernadino and Tulare Counties, California - Originally published in 1917, this important publication on California Mining has not been available for nearly a century. This publication includes important information on the early gold mines of San Bernadino and Tulare County, which were some of the first gold fields mined in California by early Spanish and Mexican miners before the 49ers came on the scene. Included are not only details on early mining methods in the area, production statistics and geological information, but also the location of the early gold mines that helped make California "The Golden State". Also included are details on the mining of other minerals such as copper, iron, lead, zinc, manganese, tungsten, vanadium, asbestos, barite, borax, cement, clay, dolomite, fluospar, gem stones, graphite, marble, salines, petroleum, stronium, talc and others. **8.5" X 11", 200 ppgs. Retail Price: $19.99**

Chromite Mining in The Klamath Mountains of California and Oregon - Unavailable since 1919, this publication was originally compiled by J.S. Diller of the United States Department of Geological Survey and includes details on the chromite mines of this area of Northern California and Southern Oregon. Also included are insights into the mineralization and other characteristics of this important mining region, as well as the location of historic mines. Also included are insights into chromite mining in Eastern Oregon and Montana. **8.5" X 11", 98 ppgs. Retail Price: $9.99**

Mines and Mining in Amador, Calaveras and Tuolumne Counties, California - Unavailable since 1915, this publication was originally compiled by William Tucker and includes details on the mines and mineral resources of this important California mining area. Included are details on the geology, history and important gold mines of the region, as well as insights into other local mineral resources such as asbestos, clay, copper, talc, limestone and others. Also included are insights into the mineralization and other characteristics of this important portion of California's Mother Lode mining region. **8.5" X 11", 198 ppgs. Retail Price: $14.99**

The Cerro Gordo Mining District of Inyo County California - Unavailable since 1963, this publication was originally compiled by the United States Department of Interior. Included are insights into the mineralization and other characteristics of this important mining region of Southern California. Topics include the mining of gold and silver in this important mining district in Inyo County, California, including details on the history, production and locations of the Cerro Gordo Mine, the Morning Star Mine, Estelle Tunnel, Charles Lease Tunnel, Ignacio, Hart, Crosscut Tunnel, Sunset, Upper Newtown, Newtown, Ella, Perseverance, Newsboy, Belmont and other silver and gold mines in the Cerro Gordo Mining District. This volume also includes important insights into the fossil record, geologic formations, faults and other aspects of economic geology in this California mining district. **8.5" X 11", 104 ppgs. Retail Price: $10.99**

Mining in Butte, Lassen, Modoc, Sutter and Tehama Counties of California - Unavailable since 1917, this publication was originally compiled by the United States Department of Interior. Included are insights into the mineralization and other characteristics of this important mining region of California. Topics include the mining of asbestos, chromite, gold, diamonds and manganese in Butte County, the mining of gold and copper in the Hayden Hill and Diamond Mountain mining districts of Lassen County, the mining of coal, salt, copper and gold in the High Grade and Winters mining districts of Modoc County, gold mining in Sutter County and the mining of gold, chromite, manganese and copper in Tehama County. This volume also includes the production records and locations of numerous mines in this important mining region. **8.5" X 11", 114 ppgs. Retail Price: $11.99**

Mines of Trinity County California - Originally published in 1965, this important publication on California Mining has not been available for nearly fifty years. This publication includes important information on mines and mining in Trinity County, California, as well insights into the mineralization and geology of this important mining area in Northern California. Included are extensive details on hardrock and placer gold mines and prospects, including charts showing the locations of these historic mines.. 8.5" X 11", 144 ppgs. **Retail Price: $12.99**

Mines of Kern County California - Originally published in 1962, this important publication on California Mining has not been available for nearly fifty years. This publication includes important information on mines and mining in Kern County, California, as well insights into the mineralization and geology of this important mining area in California. Included are extensive details on hardrock and placer gold mines and prospects, including charts showing the locations of these historic mines. 8.5" X 11", 398 ppgs. **Retail Price: $24.99**

Mines of Calaveras County California - Originally published in 1962, this important publication on California Mining has not been available for nearly fifty years. This publication includes important information on mines and mining in Calaveras County, California, as well insights into the mineralization and geology of this important mining area in Northern California. Included are extensive details on hardrock and placer gold mines and prospects, including charts showing the locations of these historic mines. 8.5" X 11", 236 ppgs. **Retail Price: $19.99**

Lode Gold Mining in Grass Valley California - Unavailable since 1940, this publication was originally compiled by the United States Department of Interior. Included are insights into the gold mineralization and other characteristics of this important mining region of Nevada County, California. This volume also includes important insights into the geologic formations, faults and other aspects of economic geology in this California mining district. Of particular interest are the fine details on many hardrock gold mines in the area, including their locations, histories, development and mineralization. Some of the mines featured include the Gold Hill Mine, Massachusetts Hill, Boundary, Peabody, Golden Center, North Star, Omaha, Lone Jack, Homeward Bound, Hartery, Wisconsin, Allison Ranch, Phoenix, Kate Hayes, W.Y.O.D., Empire, Rich Hill, Daisy Hill, Orleans, Sultana, Centennial, Conlin, Ben Franklin, Crown Point and many others. 8.5" X 11", 148 ppgs. **Retail Price: $12.99**

Alaska Mining Books

Ore Deposits of the Willow Creek Mining District, Alaska - Unavailable since 1954, this hard to find publication includes valuable insights into the Willow Creek Mining District near Hatcher Pass in Alaska. The publication includes insights into the history, geology and locations of the well known mines in the area, including the Gold Cord, Independence, Fern, Mabel, Lonesome, Snowbird, Schroff-O'Neil, High Grade, Marion Twin, Thorpe, Webfoot, Kelly-Willow, Lane, Holland and others. 8.5" X 11", 96 ppgs. **Retail Price: $9.99**

Arizona Mining Books

Mines and Mining in Northern Yuma County Arizona - Originally published in 1911, this important publication on Arizona Mining has not been available for over a hundred years. Included are rare insights into the gold, silver, copper and quicksilver mines of Yuma County, Arizona together with hard to find maps and photographs. Some of the mines and mining districts featured include the Planet Copper Mine, Mineral Hill, the Clara Consolidated Mine, Viati Mine, Copper Basin prospect, Bowman Mine, Quartz King, Billy Mack, Carnation, the Wardwell and Osbourne, Valensuella Copper, the Mariquita, Colonial Mine, the French American, the New York-Plomosa, Guadalupe, Lead Camp, Mudersbach Copper Camp, Yellow Bird, the Arizona Northern (Salome Strike), Bonanza (Harqua Hala), Golden Eagle, Hercules, Socorro and others. 8.5" X 11", 144 ppgs. **Retail Price: $11.99**

The Aravaipa and Stanley Mining Districts of Graham County Arizona - Originally published in 1925, this important publication on Arizona Mining has not been available for nearly ninety years. Included are rare insights into the gold and silver mines of these two important mining districts, together with hard to find maps. 8.5" X 11", 140 ppgs. **Retail Price: $11.99**

Gold in the Gold Basin and Lost Basin Mining Districts of Mohave County, Arizona - This volume contains rare insights into the geology and gold mineralization of the Gold Basin and Lost Basin Mining Districts of Mohave County, Arizona that will be of benefit to miners and prospectors. Also included is a significant body of information on the gold mines and prospects of this portion of Arizona. This volume is lavishly illustrated with rare photos and mining maps. 8.5" X 11", 188 ppgs. **Retail Price: $19.99**

Mines of the Jerome and Bradshaw Mountains of Arizona - This important publication on Arizona Mining has not been available for ninety years. This volume contains rare insights into the geology and ore deposits of the Jerome and Bradshaw Mountains of Arizona that will be of benefit to miners and prospectors who work those areas. Included is a significant body of information on the mines and prospects of the Verde, Black Hills, Cherry Creek, Prescott, Walker, Groom Creek, Hassayampa, Bigbug, Turkey Creek, Agua Fria, Black Canyon, Peck, Tiger, Pine Grove, Bradshaw, Tintop, Humbug and Castle Creek Mining Districts. This volume is lavishly illustrated with rare photos and mining maps. 8.5" X 11", 218 ppgs. **Retail Price: $19.99**

The Ajo Mining District of Pima County Arizona - This important publication on Arizona Mining has not been available for nearly seventy years. This volume contains rare insights into the geology and mineralization of the Ajo Mining District in Pima County, Arizona and in particular the famous New Cornelia Mine.   8.5" X 11", 126 ppgs. **Retail Price: $11.99**

Mining in the Santa Rita and Patagonia Mountains of Arizona - Originally published in 1915, this important publication on Arizona Mining has not been available for nearly a century. Included are rare insights into hundreds of gold, silver, copper and other mines in this famous Arizona mining area. Details include the locations, geology, history, production and other facts of the mines of this region. **8.5" X 11", 394 ppgs. Retail Price: $24.99**

Montana Mining Books

A History of Butte Montana: The World's Greatest Mining Camp - First published in 1900 by H.C. Freeman, this important publication sheds a bright light on one of the most important mining areas in the history of The West. Together with his insights, as well as rare photographs of the periods, Harry Freeman describes Butte and its vicinity from its early beginnings, right up to its flush years when copper flowed from its mines like a river. At the time of publication, Butte, Montana was known worldwide as "The Richest Mining Spot On Earth" and produced not only vast amounts of copper, but also silver, gold and other metals from its mines. Freeman illustrates, with great detail, the most important mines in the vicinity of Butte, providing rare details on their owners, their history and most importantly, how the mines operated and how their treasures were extracted. Of particular interest are the dozens of rare photographs that depict mines such as the famous Anaconda, the Silver Bow, the Smoke House, Moose, Paulin, Buffalo, Little Minah, the Mountain Consolidated, West Greyrock, Cora, the Green Mountain, Diamond, Bell, Parnell, the Neversweat, Nipper, Original and many others.   **8.5" X 11", 142 ppgs. Retail Price: $12.99**

The Butte Mining District of Montana - This important publication on Montana Mining has not been available for over a century. Included are rare insights into the gold, copper and silver mines of Butte, Montana together with hard to find maps and photographs. Some of the topics include the early history of gold, silver and copper mining in the Butte area, insight into the geology of its mining areas, the local distribution of gold, silver and copper ores, as well their composition and how to identify them. Also included are detailed facts about the mines in the Butte Mining District, including the famous Anaconda Mine, Gagnon, Parrot, Blue Vein, Moscow, Poulin, Stella, Buffalo, Green Mountain, Wake Up Jim, the Diamond-Bell Group, Mountain Consolidated, East Greyrock, West Greyrock, Snowball, Corra, Speculator, Adirondack, Miners Union, the Jessie-Edith May Group, Otisco, Iduna, Colorado, Lizzie, Cambers, Anderson, Hesperus, Preferencia and dozens of others. **8.5" X 11", 298 ppgs. Retail Price: $24.99**

Mines of the Helena Mining Region of Montana - This important publication on Montana Mining has not been available for over a century. Included are rare insights into the gold, copper and silver mines of the vicinity of Helena, Montana, including the Marysville Mining District, Elliston Mining District, Rimini Mining District, Helena Mining District, Clancy Mining District, Wickes Mining District, Boulder and Basin Mining Districts and the Elkhorn Mining District. Some of the topics include the early history of gold, silver and copper mining in the Helena area, insight into the geology of its mining areas, the local distribution of gold, silver and copper ores, as well their composition and how to identify them. Also included are detailed facts, history, geology and locations of over one hundred gold, silver and copper mines in the area . **8.5" X 11", 162 ppgs, Retail Price: $14.99**

Mines and Geology of the Garnet Range of Montana - This important publication on Montana Mining has not been available for over a century. Included are rare insights into the gold, copper and silver mines of the vicinity of this important mining area of Montana. Some of the topics include the early history of gold, silver and copper mining in the Garnet Mountains, insight into the geology of its mining areas, the local distribution of gold, silver and copper ores, as well their composition and how to identify them. Also included are detailed facts, history, geology and locations of numerous gold, silver and copper mines in the area . **8.5" X 11", 100 ppgs, Retail Price: $11.99**

Mines and Geology of the Philipsburg Quadrangle of Montana - This important publication on Montana Mining has not been available for over a century. Included are rare insights into the gold, copper and silver mines of the vicinity of this important mining area of Montana. Some of the topics include the early history of gold, silver and copper mining in the Philipsburg Quadrangle, insight into the geology of its mining areas, the local distribution of gold, silver and copper ores, as well their composition and how to identify them. Also included are detailed facts, history, geology and locations of over one hundred gold, silver and copper mines in the area **8.5" X 11", 290 ppgs, Retail Price: $24.99**

Geology of the Marysville Mining District of Montana - Included are rare insights into the mining geology of the Marysville Mining District. Some of the topics include the early history of gold, silver and copper mining in the area, insight into the geology of its mining areas, the local distribution of gold, silver and copper ores, as well their composition and how to identify them. Also included are detailed facts, history, geology and locations of gold, silver and copper mines in the area   **8.5" X 11", 198 ppgs, Retail Price: $19.99**

<u>**The Geology and Mines of Northern Idaho and North Western Montana**</u>

See listing under Idaho.

Nevada Mining Books

<u>**The Bull Frog Mining District of Nevada**</u> - Unavailable since 1910, this publication was originally compiled by the United States Department of Interior. This volume also includes important insights into the geologic formations, faults and other aspects of economic geology in this Nevada mining district. Of particular interest are the fine details on many mines in the area, including their locations, histories, development and mineralization. Some of the mines featured include the National Bank Mine, Providence, Gibraltor, Tramps, Denver, Original Bullfrog, Gold Bar, Mayflower, Homestake-King and other mines and prospects. **8.5" X 11", 152 ppgs, Retail Price: $14.99**

Colorado Mining Books

<u>**Ores of The Leadville Mining District**</u> - Unavailable since 1926, this publication was originally compiled by the United States Department of Interior. This volume also includes important insights into the ores and mineralization of the Leadville Mining District in Colorado. Topics include historic ore prospecting methods, local geology, insights into ore veins and stockworks, the local trend and distribution of ore channels, reverse faults, shattered rock above replacement ore bodies, mineral enrichment in oxidized and sulphide zones and more. **8.5" X 11", 66 ppgs, Retail Price: $8.99**

<u>**Mining in Colorado**</u> - Unavailable since 1926, this publication was originally compiled by the United States Department of Interior. This volume also includes important insights into the mining history of Colorado from its early beginnings in the 1850's right up to the mid 1920's. Not only is Colorado's gold mining heritage included, but also its silver, copper, lead and zinc mining industry. Each mining area is treated separately, detailing the development of Colorado's mines on a county by county basis. **8.5" X 11", 284 ppgs, Retail Price: $19.99**

<u>Gold Mining in Gilpin County Colorado</u> - Unavailable since 1876, this publication was originally compiled by the Register Steam Printing House of Central City, Colorado. A rare glimpse at the gold mining history and early mines of Gilpin County, Colorado from their first discovery in the 1850's up to the "flush years" of the mid 1870's. Of particular interest is the history of the discovery of gold in Gilpin County and details about the men who made those first strikes. Special focus is given to the early gold mines and first mining districts of the area, many of which are not detailed in other books on Colorado's gold mining history. **8.5" X 11", 156 ppgs, Retail Price: $12.99**

<u>Mining in the Gold Brick Mining District of Colorado</u> - Important insights into the history of the Gold Brick Mining District, as well as its local geography and economic geology. Also included are the histories and locations of historic mines in this important Colorado Mining District, including the Cortland, Carter, Raymond, Gold Links, Sacramento, Bassick, Sandy Hook, Chronicle, Grand Prize, Chloride, Granite Mountain, Lucille, Gray Mountain, Hilltop, Maggie Mitchell, Silver Islet, Revenue, Roosevelt, Carbonate King and others. In addition to hardrock mining, are also included are details on gold placer mining in this portion of Colorado. **8.5" X 11", 140 ppgs, Retail Price: $12.99**

Washington Mining Books

<u>**The Republic Mining District of Washington**</u> - Unavailable since 1910, this important publication was originally published by the Washington Geologic Survey and has been unavailable for a century. Topics include the geology, rock formations and the formation of ore deposits in this important mining area of Washington State. Also included are hard to find details on the geology, history and locations of dozens of mines in the area. Some of the mines featured include the New Republic Mine, Ben Hur, Morning Glory, the South Republic Mine, Quilp, Surprise, Black Tail, Lone Pine, San Poil, Mountain Lion, Tom Thumb, Elcaliph and many others. **8.5" X 11", 94 ppgs, Retail Price: $10.99**

Wyoming Mining Books

<u>**Mining in the Laramie Basin of Wyoming**</u> - Unavailable since 1909, this publication was originally compiled by the United States Department of Interior. Also included are insights into the mineralization and other characteristics of this important mining region, especially in regards to coal, limestone, gypsum, bentonite clay, cement, sand, clay and copper. **8.5" X 11", 104 ppgs, Retail Price: $11.99**

More Mining Books

Prospecting and Developing A Small Mine - Topics covered include the classification of varying ores, how to take a proper ore sample, the proper reduction of ore samples, alluvial sampling, how to understand geology as it is applied to prospecting and mining, prospecting procedures, methods of ore treatment, the application of drilling and blasting in a small mine and other topics that the small scale miner will find of benefit. **8.5" X 11", 112 ppgs, Retail Price: $11.99**

Timbering For Small Underground Mines - Topics covered include the selection of caps and posts, the treatment of mine timbers, how to install mine timbers, repairing damaged timbers, use of drift supports, headboards, squeeze sets, ore chute construction, mine cribbing, square set timbering methods, the use of steel and concrete sets and other topics that the small underground miner will find of benefit. This volume also includes twenty eight illustrations depicting the proper construction of mine timbering and support systems that greatly enhance the practical usability of the information contained in this small book. **8.5" X 11", 88 ppgs. Retail Price: $10.99**

Timbering and Mining - A classic mining publication on Hard Rock Mining by W.H. Storms. Unavailable since 1909, this rare publication provides an in depth look at American methods of underground mine timbering and mining methods. Topics include the selection and preservation of mine timbers, drifting and drift sets, driving in running ground, structural steel in mine workings, timbering drifts in gravel mines, timbering methods for driving shafts, positioning drill holes in shafts, timbering stations at shafts, drainage, mining large ore bodies by means of open cuts or by the "Glory Hole" system, stoping out ore in flat or low lying veins, use of the "Caving System", stoping in swelling ground, how to stope out large ore bodies, Square Set timbering on the Comstock and its modifications by California miners, the construction of ore chutes, stoping ore bodies by use of the "Block System", how to work dangerous ground, information on the "Delprat System" of stoping without mine timbers, construction and use of headframes and much more. This volume provides a reference into not only practical methods of mining and timbering that may be employed in narrow vein mining by small miners today, but also rare insights into how mines were being worked at the turn of the 19th Century. **8.5" X 11", 288 ppgs. Retail Price: $24.99**

A Study of Ore Deposits For The Practical Miner - Mining historian Kerby Jackson introduces us to a classic mining publication on ore deposits by J.P. Wallace. First published in 1908, it has been unavailable for over a century. Included are important insights into the properties of minerals and their identification, on the occurrence and origin of gold, on gold alloys, insights into gold bearing sulfides such as pyrites and arsenopyrites, on gold bearing vanadium, gold and silver tellurides, lead and mercury tellurides, on silver ores, platinum and iridium, mercury ores, copper ores, lead ores, zinc ores, iron ores, chromium ores, manganese ores, nickel ores, tin ores, tungsten ores and others. Also included are facts regarding rock forming minerals, their composition and occurrences, on igneous, sedimentary, metamorphic and intrusive rocks, as well as how they are geologically disturbed by dikes, flows and faults, as well as the effects of these geologic actions and why they are important to the miner. Written specifically with the common miner and prospector in mind, the book will help to unlock the earth's hidden wealth for you and is written in a simple and concise language that anyone can understand. **8.5" X 11", 366 ppgs. Retail Price: $24.99**

Mine Drainage - Unavailable since 1896, this rare publication provides an in depth look at American methods of underground mine drainage and mining pump systems. This volume provides a reference into not only practical methods of mining drainage that may be employed in narrow vein mining by small miners today, but also rare insights into how mines were being worked at the turn of the 19th Century. **8.5" X 11", 218 ppgs. Retail Price: $24.99**

Fire Assaying Gold, Silver and Lead Ores - Unavailable since 1907, this important publication was originally published by the Mining and Scientific Press and was designed to introduce miners and prospectors of gold, silver and lead to the art of fire assaying. Topics include the fire assaying of ores and products containing gold, silver and lead; the sampling and preparation of ore for an assay; care of the assay office, assay furnaces; crucibles and scorifiers; assay balances; metallic ores; scorification assays; cupelling; parting' crucible assays, the roasting of ores and more. This classic provides a time honored method of assaying put forward in a clear, concise and easy to understand language that will make it a benefit to even beginners. **8.5" X 11", 96 ppgs. Retail Price: $11.99**

Methods of Mine Timbering - Originally published in 1896, this important publication on mining engineering has not been available for nearly a century. Included are rare insights into historical methods of timbering structural support that were used in underground metal mines during the California that still have a practical application for the small scale hardrock miner of today. **8.5" X 11", 94 ppgs. Retail Price: $10.99**

www.ingramcontent.com/pod-product-compliance
Lightning Source LLC
Chambersburg PA
CBHW080257180526
45167CB00006B/2560